Key Questions in Animal Behaviour and Welfare:
A Study and Revision Guide

Key Questions in Animal Behaviour and Welfare: A Study and Revision Guide

Paul A. Rees *BSc(Hons), LLM, PhD, CertEd*
Formerly Senior Lecturer
School of Science, Engineering and Environment, University of Salford,
United Kingdom

CABI is a trading name of CAB International

CABI
Nosworthy Way
Wallingford
Oxfordshire OX10 8DE
UK

Tel: +44 (0)1491 832111
E-mail: info@cabi.org
Website: www.cabi.org

CABI
WeWork
One Lincoln St
24th Floor
Boston, MA 02111
USA

Tel: +1 (617)682-9015
E-mail: cabi-nao@cabi.org

A catalogue record for this book is available from the British Library, London, UK.

References to Internet websites (URLs) were accurate at the time of writing.

ISBN-13: 978 1 78924 897 5 (paperback)
       978 1 78924 898 2 (ePDF)
       978 1 78924 899 9 (ePub)

DOI: 10.1079/9781789248975.0000

Commissioning Editor: Ward Cooper
Editorial Assistant: Emma McCann
Production Editor: James Bishop

Typeset by SPi, Pondicherry, India
Printed and bound in the UK by CPI Group (UK) Ltd, Croydon, CR0 4YY

For Elliot

Contents

Contents

About the Author

Paul Rees was a senior lecturer in the School of Science, Engineering and Environment at the University of Salford, United Kingdom, for 22 years until his retirement in 2020. He holds a BSc(Hons) in Environmental Biology from the University of Liverpool, a PhD in animal ecology and behaviour from the University of Bradford and an LLM in Environmental Law from the University of Central Lancashire. Paul previously lectured at three Further Education Colleges and a Higher Education College in the United Kingdom, and trained biology teachers at the Sokoto College of Education in Nigeria. He has taught from GCE 'O'/GCSE level to MSc level and has been an external examiner for a range of taught programmes, from Higher National Diploma to MSc level, at six British universities. Paul has published papers on mammal behaviour and ecology, wildlife law, and the role of zoos in conservation, along with ten textbooks concerned with ecology, zoo biology, wildlife law and elephants. Paul is the author of four other titles in CABI's *Key Questions* series:

Key Questions in Ecology
Key Questions in Applied Ecology and Conservation
Key Questions in Biodiversity
Key Questions in Zoo and Aquarium Studies

Preface

This book is intended to introduce college and university students – and anyone interested in the subjects – to animal behaviour and animal welfare. I have grouped these disciplines together because they are cognate subjects and many of the topics covered are relevant to both, such as neurophysiology, the measurement and recording of behaviour, and cognition.

The book offers the reader the opportunity to answer 600 multiple choice questions on a wide range of topics including the history of animal behaviour and animal welfare science, neurophysiology, learning, territoriality, animal navigation, migration, cognition and communication, behavioural ecology, social behaviour, animal exploitation and animal rights. It also considers methods of studying behaviour and the ethical and legal issues relating to animal welfare and experimentation on animals.

In recent years many college and university courses have been adopting multiple-choice questions (MCQs) as a method of assessment. In my experience students have a positive attitude to this type of assessment and enjoy practising answering MCQs in class. This trend, and the sudden switch to online teaching and learning at many institutions around the world as a result of the social distancing necessary to slow the spread of the COVID-19 pandemic, have created a unique opportunity for books of MCQs to add value to many science courses.

It is not possible to cover every aspect of animal behaviour and animal welfare in a book of this nature and size. However, I hope that using it will encourage you to find out more about those topics that interest you and improve your overall understanding of the challenges and rewards of studying animal behaviour and the importance of achieving good animal welfare in all of our interactions with other animal species.

Preface

Acknowledgements

I am grateful to Ward Cooper (Commissioning Editor) and his colleagues at CABI for their encouragement and support during the production of this book.

Figure 4.4 was taken by my daughter, Clara Clark, and I am very grateful to her for allowing me to publish it here. Figure 6.3 was taken by Kathryn Page as part of a project being conducted by Helen Chambers (University of Salford) on bear cognition, and I am most grateful to her for allowing me to reproduce this. Figure 5.4 is adapted from a figure in Krebs (1982).

Some of the questions in this book are based on those I previously set as part of the assessment for the BSc(Hons) in Wildlife Conservation with Zoo Biology and the BSc(Hons) in Wildlife and Practical Conservation at the University of Salford, and I am grateful to those students who have unwittingly tested them for me.

Finally, I would like to thank my wife Katy for understanding that I prefer to write than decorate the house or manage the garden, and for waiting patiently for me to engage with these activities (eventually).

How to use this book

The questions in each chapter are divided into three levels: foundation, intermediate, and advanced. These levels are not intended to reflect any particular curriculum but rather general levels of difficulty, and should not be taken too seriously. Knowledge of basic facts are dealt with at the foundation level while the intermediate level and advanced levels contain questions involving more obscure facts and concepts. However, there is some variation between chapters as not all of the areas covered lend themselves to this approach. Students are advised to check the syllabuses they are following in detail before relying too much on this book as a preparation for specific exams.

Students are encouraged to complete a whole chapter – or at least a complete section (foundation, intermediate or advanced) – before looking at the answers. This is because the explanations for some answers may assist in selecting the correct answer to subsequent questions, although I have tried to avoid this where possible. The order in which the chapters are attempted does not really matter because each is about a distinct topic. However, within any chapter you are advised to attempt the foundation questions first, followed by the intermediate questions and finally the advanced questions.

1 Foundations and History of Animal Behaviour and Welfare

This chapter contains questions about the origins of the studies of animal behaviour and animal welfare, their historical development and the roles of important historical figures and their publications.

Foundation

1.1f Which of the following is the most accurate definition of the term ethology?

  a. It is the study of animal behaviour

  b. It is the study of animal behaviour in the laboratory

  c. It is the study of the behaviour of animals in their natural environment

  d. It is an alternate to the term 'experimental psychology'

1.2f Which of the following are considered to be the founders of modern ethology?

  a. Patrick Bateson and Richard Dawkins

  b. Konrad Lorenz and Nikolaas Tinbergen

  c. Marian Dawkins and Konrad Lorenz

  d. Nikolaas Tinbergen and John Krebs

© Paul A. Rees 2022. *Key Questions in Animal Behaviour and Welfare: A Study and Revision Guide* (P.A. Rees)
DOI: 10.1079/9781789248975.0001

1.3f   It is possible to study and manipulate animal behaviour without a complete understanding of the underlying mechanisms involved. For example, an ape can be trained to cooperate with medical procedures such as providing a blood sample without the trainer or the veterinary surgeon involved understanding the physiological mechanisms involved (Fig. 1.1). Such an approach treats the animal as a

   a. 'blue box'

   b. 'white box'

   c. 'red box'

   d. 'black box'

Fig. 1.1.

1.4f   Professor Timothy Clutton-Brock, of the Department of Zoology of the University of Cambridge, is famous for his long-term field studies of

   a. red deer and meerkats

   b. grey wolves and capuchin monkeys

   c. red foxes and Asian elephants

   d. mountain gorillas and European badgers

1.5f   An animal's behaviour develops under the influence of its

   a. genes only

   b. environment only

   c. genes and physical environment

   d. genes and physical and biological environment

1.6f Who, in a book published in 1859, considered whether the behavioural traits of animals could evolve by natural selection?

a. Richard Owen

b. Charles Darwin

c. John Wray (Ray)

d. Georges Cuvier

1.7f The school of psychology which explains behaviour only in terms of observable stimuli, muscular movements and glandular secretions is known as

a. behaviourism

b. cognitive psychology

c. gestalt psychology

d. psychoanalysis

1.8f The science that attempts to explain social behaviour in terms of evolution is called

a. sociology

b. sociobiology

c. biosociology

d. social psychology

1.9f In 1932 Sir Solly Zuckerman published a book entitled *The Social Life of*

a. *Monkeys and Apes*

b. *Elephants*

c. *Mammals*

d. *Birds*

1.10f *The Naked Ape* and *The Human Zoo* are books written about human behaviour by the zoologist

a. Edward O. Wilson

b. Hans Kruuk

c. Richard Dawkins

d. Desmond Morris

1.11f **The study of the similarities and differences in the behaviour of different types of animals – especially the extent to which animal behaviour can help us to understand human behaviour – is known as**

    a. comparative neurophysiology

    b. comparative psychology

    c. comparative zoology

    d. comparative behaviour

1.12f **Who argued for better treatment for animals in his book *An Introduction to the Principles of Morals and Legislation* published in 1780?**

    a. Jeremy Bentham

    b. Emmanuel Kant

    c. Lewis Gompertz

    d. René Descartes

1.13f **Laws concerned with preventing and punishing animal cruelty are likely to include cruelty resulting from which of the following?**

    i. Acts of neglect

    ii. Omissions

    iii. Over-working

    iv. Exposure to excessive heat or cold

    v. Mutilation

    a. i, iii and v

    b. ii, iii, iv and v

    c. i, iii, iv and v

    d. i, ii, iii, iv and v

1.14f **Which of the following journals publish papers on the behaviour of animals in zoos?**

    i. *Animal Welfare*

    ii. *Zoo Biology*

iii. *Journal of Applied Animal Welfare Science*

iv. *Applied Animal Behaviour Science*

a. i and ii

b. ii, iii and iv

c. i, ii and iii

d. All of them

1.15f Ivan Petrovich Pavlov was

a. an American psychologist who studied social behaviour in monkeys

b. a Polish behaviourist who studied problem solving in rats

c. a Russian physiologist who studied conditioning

d. an Austrian ethologist who studied learning in birds

1.16f The Edward Grey Institute is part of the Zoology Department of the University of Oxford and is famous for studies of the ecology and behaviour of

a. mammals

b. primates

c. birds

d. reptiles

1.17f Complete the following sentence with one of the terms listed below: 'The welfare of an animal may be defined as its state as regards its attempts to cope with its ....'.

a. feelings

b. environment

c. perceptions

d. sensations

1.18f *Blackfish* was a television documentary that was important in changing public attitudes to

a. angling

b. fish farming

   c. keeping orcas in captivity

   d. the sea fishing industry

1.19f The electroencephalograph (EEG) was invented in the 1920s. It is a machine for recording

   a. brain activity

   b. hormonal activity

   c. muscle activity

   d. animal behaviour

1.20f David Lack discovered that robins (*Erithacus rubecula*) sing to defend their territory. Ernest Neal was an expert on badgers (*Meles meles*) and produced the first film of their nocturnal activities. At the beginning of their studies of these animals they were both

   a. university lecturers

   b. government scientists

   c. engineers

   d. schoolteachers

Intermediate

1.1i Which of the following was not a major figure in the development of behaviourism?

   a. John B. Watson

   b. Burrhus F. Skinner

   c. Ivan Pavlov

   d. Jean Piaget

1.2i Which of the following scientists were awarded the Nobel Prize in Physiology or Medicine in 1973 'for their discoveries concerning organization and elicitation of individual and social behaviour patterns?'

   a. George Schaller, Jane Goodall and Konrad Lorenz

   b. Konrad Lorenz, Nikolaas Tinbergen and Karl von Frisch

   c. Frans de Waal, Jane Goodall and Nikolaas Tinbergen

   d. Karl von Frisch, George Schaller and Konrad Lorenz

1.3i **A long-term study of the chimpanzees (*Pan troglodytes*) in the Gombi Stream National Park, Tanzania was started in the 1960s by**

   a. Robert Hinde

   b. John Napier

   c. Jane Goodall

   d. Dian Fossey

1.4i **The 'five freedoms' used to assess animal welfare were established in a report by an expert committee in Britain led by**

   a. Prof. E. L. Brampton

   b. Prof. F. W. R. Brambell

   c. Prof. M. E. Bramwell

   d. Prof. S. D. T. Branston

1.5i **Cynthia Moss has made a long-term study of**

   a. African elephants (*Loxodonta africana*) in Amboseli National Park, Kenya

   b. Asian elephants (*Elephas maximus*) in Kaziranga National Park, India

   c. American bison (*Bison bison*) in the Badlands National Park, South Dakota, United States

   d. giraffes (*Giraffa camelopardalis*) in the Serengeti National Park, Tanzania

1.6i **The first scientific journal to be published on the subject of animal behaviour was published in**

   a. the United Kingdom

   b. Austria

   c. Germany

   d. the United States

1.7i In 1676 John Wray (Ray) published a scientific text on the study of instinctive behaviour in

a. rats

b. birds

c. dogs

d. cats

1.8i Harry F. Harlow is best known for his studies of

a. language in chimpanzees

b. imprinting in geese

c. operant conditioning in dogs

d. maternal separation in monkeys

1.9i Which of the following scientists could not be considered an ethologist?

a. Nikolaas Tinbergen

b. Konrad Lorenz

c. Karl von Frisch

d. Burrhus F. Skinner

1.10i Who was the author of *The Herring Gull's World*?

a. Nikolaas Tinbergen

b. Desmond Morris

c. David Lack

d. Timothy Birkhead

1.11i The behaviourists explained animal behaviour almost exclusively in terms of

a. evolution

b. learning

c. innate behaviour

d. neurophysiology

1.12i **The publications in Table 1.1 have been important in the development of animal welfare and animal rights. Which of the columns A–D lists the works in the correct chronological order by date of publication (earliest first)?**

Table 1.1

	A	B	C	D
Oldest	Animal Machines	Mental Evolution in Animals	Animal Machines	The Question of Animal Awareness
	The Question of Animal Awareness	Animal Machines	Mental Evolution in Animals	Animal Machines
	The Case for Animal Rights	The Question of Animal Awareness	The Question of Animal Awareness	Mental Evolution in Animals
Most recent	Mental Evolution in Animals	The Case for Animal Rights	The Case for Animal Rights	The Case for Animal Rights

    a. A

    b. B

    c. C

    d. D

1.13i **Who wrote a book entitled *In the Shadow of Man*?**

    a. George Schaller

    b. John Napier

    c. Jane Goodall

    d. Dian Fossey

1.14i **A long-term experiment conducted in Siberia studying the process of domestication examined the inheritance of tameness in the**

    a. silver fox (*Vulpes vulpes*)

    b. grey wolf (*Canis lupus*)

    c. sable (*Martes zibellina*)

    d. pine marten (*Martes martes*)

1.15i Who published a paper in 1914 entitled *The courtship habits of the Great Crested Grebe* (Podiceps cristatus); *with an addition to the Theory of Sexual Selection*?

   a. Julian Huxley

   b. Edward Grey

   c. David Lack

   d. Frank Fraser Darling

1.16i A study of pet ownership in 700 households in Ontario, Canada (Leslie *et al.*, 1994) found that the main reason that animals were kept was for

   a. the benefit of children

   b. companionship

   c. love and affection

   d. entertainment

1.17i In 1925 Wolfgang Köhler published a book entitled

   a. *The Mentality of Elephants*

   b. *The Mentality of Dolphins*

   c. *The Mentality of Dogs*

   d. *The Mentality of Apes*

1.18i John Bowlby applied ethological methods to the study of

   a. ape communication

   b. child development

   c. bird migration

   d. animals in zoos

1.19i Which of the following conducted a detailed study of the social structure and behaviour of African elephants (*Loxodonta africana*) in Lake Manyara National Park, Tanzania?

   a. Raman Sukumar

   b. Iain Douglas-Hamilton

c. Ian Redmond

d. Harvey Croze

1.20i **Who wrote the books *Enriching Animal Lives* and *Behavioral Enrichment in the Zoo*?**

a. Terry Maple

b. Georgia Mason

c. Desmond Morris

d. Hal Markowitz

Advanced

1.1a **In 1964 the British animal welfare activist Ruth Harrison published a book entitled**

a. *Animal Welfare*

b. *Animal Machines*

c. *Animal Cruelty*

d. *Animal Automatons*

1.2a **The world's first professor of animal welfare was**

a. Donald Broom

b. John Webster

c. David Mellor

d. Marian Stamp Dawkins

1.3a **The first Chair in Animal Welfare was created at the University of**

a. Oxford

b. Cambridge

c. London

d. Edinburgh

1.4a **Who wrote the book *On Aggression*?**

   a. Marian Dawkins

   b. Desmond Morris

   c. Konrad Lorenz

   d. Richard Dawkins

1.5a **What is the name of the prestigious animal behaviour research organisation located in Konstanz, Germany?**

   a. Max Planck Institute of Animal Behaviour

   b. Paul Ehrlich Institute of Animal Behaviour

   c. Alexander von Humboldt Institute of Animal Behaviour

   d. Max Delbrück Institute of Animal Behaviour

1.6a **Which of the following is not associated with early uses of the term 'ethology' with its current meaning?**

   a. William Morton Wheeler

   b. Isidore Geoffrey Saint-Hilaire

   c. Oskar Heinroth

   d. Ernst Haeckel

1.7a **Who made a classic field study of the behaviour and ecology of African lions (*Panthera leo*) and published his work in 1976 in a book entitled *The Serengeti Lion*?**

   a. Gary Packer

   b. George Schaller

   c. L. David Mech

   d. Richard Estes

1.8a **Who founded the Animal Behaviour Research Group at the University of Oxford?**

   a. Desmond Morris

   b. Robert Hinde

   c. David Lack

   d. Nikolaas Tinbergen

1.9a **The first legal attempts to prevent cruelty to animals in the United Kingdom focussed on animals that were**

a. mammals

b. sentient

c. primates

d. intelligent

1.10a **Who was the first person to use the term 'evolutionarily stable strategy'?**

a. Geoffrey Parker

b. John Maynard Smith

c. John Krebs

d. Richard Lewontin

1.11a **Why is the animal performing the behaviour? How did the behaviour evolve? What causes the behaviour to be performed? How has the behaviour developed during the individual's lifetime? These four questions laid the foundation for how animal behaviour research should be performed and were formulated by**

a. Karl von Frisch

b. Burrhus F. Skinner

c. Nikolaas Tinbergen

d. Konrad Lorenz

1.12a **Behavioural ecologists are also known as**

a. adaptationists

b. modificationists

c. transformationists

d. integrationists

1.13a **The science of sociobiology gained major recognition as the result of a book published by**

a. Robert MacArthur

b. Stephen Jay Gould

    c. Edward O. Wilson

    d. Richard Lewontin

1.14a The study of the adaptations that animals have made to their behaviour as a result of selective pressures in the environment is called

    a. cultural ethology

    b. behavioural ecology

    c. cognitive ethology

    d. behaviourism

1.15a Which of the following scientists is renowned for his work on reciprocal altruism, parental investment and parent-offspring conflict?

    a. Robert Hinde

    b. Patrick Bateson

    c. Julian Huxley

    d. Robert Trivers

1.16a The National Primate Research Center at Emory University in Atlanta, Georgia, United States, is named after

    a. Dian Fossey

    b. Robert Yerkes

    c. Frans de Waal

    d. Jane Goodall

1.17a Who was the author of *Evolution and the Theory of Games* published in 1982?

    a. Richard Lewontin

    b. Robert May

    c. John Maynard Smith

    d. Eric Charnov

1.18a The first veterinary school was established in 1762 in

    a. Lyon, France

    b. London, England

    c. Göttingen, Germany

    d. Skara, Sweden

1.19a Who published a paper entitled *Love in Infant Monkeys* in 1959?

    a. Harry Harlow

    b. William Thorpe

    c. John Bowlby

    d. Robert Yerkes

1.20a The school of psychology that attempts to understand learning, perception and other elements of mental life as an organised whole is

    a. cognitive psychology

    b. humanistic psychology

    c. kleinian psychology

    d. gestalt psychology

2 Basic Concepts and Mechanisms in Behaviour

This chapter contains questions about taxes, kineses, instinct, reaction chains, activity patterns and rhythms, feeding, and the role of genetics in animal behaviour.

Foundation

2.1f Some insects will move towards a unidirectional source of light. This behaviour is known as

a. positive phototaxis

b. negative phototropism

c. negative phototaxis

d. positive phototropism

2.2f When an animal is repeatedly exposed to a harmless stimulus – for example a noise – it may stop responding to it. This phenomenon is called

a. a reflex reaction

b. habituation

c. a chain reaction

d. conditioning

© Paul A. Rees 2022. *Key Questions in Animal Behaviour and Welfare: A Study and Revision Guide* (P.A. Rees)
DOI: 10.1079/9781789248975.0002

2.3f The total effect an individual has on passing on its genes by producing its own offspring and by providing assistance to close relatives to produce and care for their offspring is called its

    a. inclusive fitness

    b. relative fitness

    c. summative fitness

    d. adaptive fitness

2.4f An animal that is neophobic is

    a. attracted to novel foods

    b. attracted to novel things

    c. afraid of novel things

    d. afraid of humans

2.5f The evolutionary trend towards the concentration of the sense organs at the anterior end of the body is called

    a. centralisation

    b. integration

    c. systematisation

    d. cephalisation

2.6f When Tinbergen and Lorenz moved the model shown in Fig. 2.1 through the air over a pen containing young geese, which direction of movement evoked fear reactions?

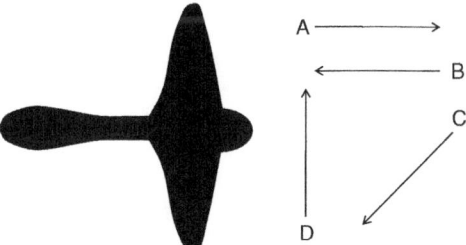

Fig. 2.1.

    a. A

    b. B

    c. C

    d. D

2.7f **A downward movement of an animal in response to the direction of gravity is known as**

a. negative geotaxis

b. positive geotaxis

c. negative geokinesis

d. positive geokinesis

2.8f **Many species of bird exhibit elaborate courtship displays. Which of the following behaviours is not an element in the courtship of the great crested grebe (*Podiceps cristatus*) (Fig. 2.2)?**

a. the ghost position

b. the penguin dance

c. the cat position

d. the pondweed dance

Fig. 2.2.

2.9f The black-throated honey guide (*Indicator indicator*) leads a honey badger (*Mellivora capensis*) to the hives of wild bees. The badger opens the hive and feeds on the honeycomb. The bird then gains access to the bee larvae and wax inside the hive. This mutually beneficial relationship is known as

    a. symbiosis

    b. altruism

    c. collaboration

    d. synergism

2.10f The killing of siblings is known as

    a. matricide

    b. kainism

    c. patricide

    d. fatalism

2.11f Which of the following species is myrmecophagous ?

    a. Aardvark (*Orycteropus afer*)

    b. Indo-Pacific bottlenose dolphin (*Tursiops aduncus*)

    c. Capybara (*Hydrochoerus hydrochaeris*)

    d. Platypus (*Ornithorhyncus anatinus*)

2.12f Black-headed gulls (*Chroicocephalus ridibundus*), the California ground squirrel (*Otospermophilus beecheyi*) and the barn swallow (*Hirundo rustica*) all live in colonies and collectively attack potential predators in a behaviour known as 'mobbing'. This is a behavioural example of

    a. divergent evolution

    b. convergent evolution

    c. adaptive radiation

    d. coevolution

2.13f **The ability of an individual to transmit genes to the next generation by producing and rearing its own offspring is its**

a. indirect fitness

b. inclusive fitness

c. direct fitness

d. complete fitness

2.14f **When an animal shows a greater response to an exaggerated stimulus than to a normal stimulus, the stimulus is referred to as**

a. a sign stimulus

b. a superstimulus

c. a substimulus

d. a major stimulus

2.15f **The manner in which an animal appreciates the world in which it lives through its senses is known as**

a. perception

b. sensation

c. cognition

d. consciousness

2.16f **A behaviour pattern in which an animal changes its activity level in response to an alteration in stimulus intensity is called a**

a. tropism

b. taxis

c. kinesis

d. tropotaxis

2.17f **An animal described as matutinal is active**

a. at dusk

b. throughout the day

c. at night

d. at dawn or early morning

2.18f **Some animals – such as some insects, reptiles and amphibians – aestivate in order to**

a. avoid predators

b. avoid periods of low temperature

c. avoid periods of high temperature

d. increase their metabolic rate

2.19f **A pollinating insect, such as a honeybee, that limits its visits to one flower type – even in the presence of other flower types – is said to exhibit**

a. flower loyalty

b. flower devotion

c. flower fidelity

d. flower allegiance

2.20f. **Figure 2.3 is a stylised representation of the vertebrate nervous system. Match the correct processes and structures listed in Table 2.1 with the labels P, Q, R and S.**

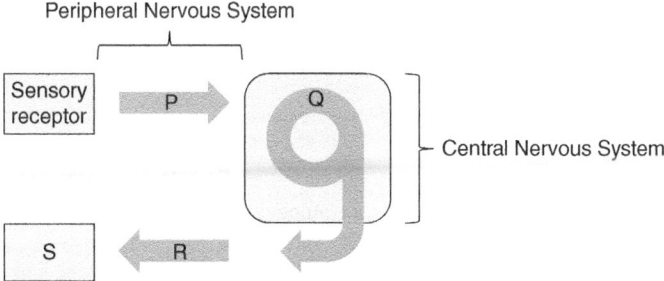

Fig. 2.3.

Table 2.1

Label	A	B	C	D
P	Sensory input	Sensory input	Motor output	Sensory input
Q	Effector organ	Integration	Integration	Integration
R	Motor output	Effector organ	Effector organ	Motor output
S	Integration	Motor output	Sensory input	Effector organ

a. A

b. B

c. C

d. D

Intermediate

2.1i **Aposematism is an advertising method used by some animal species to**

a. attract a mate

b. protect themselves from predators

c. display strength to conspecific competitors

d. attract food organisms

2.2i **The series of stimuli and responses (Fig. 2.4) observed in the courtship of three-spined sticklebacks (*Gasteroteus aculeatus*) is called**

a. a courtship reaction

b. a reaction chain

c. a sign stimulus reaction

d. a ritualisation

Male	Female
	Appears
Performs zig-zag dance	
	Courts the male – lifts her head and tail and displays her swollen abdomen
Leads the female	
	Follows the male
Shows the female the nest entrance	
	Enters the nest
Trembles	
	Spawns
Fertilises eggs	

Fig. 2.4.

2.3i   A complex behavioural sequence that is indivisible and runs to completion once initiated in response to a releaser is called

a. a fixed action pattern

b. a reflex

c. an agonistic behaviour

d. a conditioned response

2.4i   Which of the following mammals does not have a harem mating system?

a. Plains zebra (*Equus quagga*)

b. Northern elephant seal (*Mirounga angustirostris*)

c. Hamadryas baboon (*Papio hamadryas*)

d. Baird's tapir (*Tapirus bairdii*)

2.5i   Which of the following colours most commonly acts as a sign stimulus in animals?

a. Blue

b. Yellow

c. Red

d. Green

2.6i   The processes of thinking, knowing, remembering, problem-solving and making judgements are collectively termed

a. cognition

b. sentience

c. consciousness

d. self-awareness

2.7i   A behaviour that spreads by means of imitation from individual to individual is called a

a. deme

b. meme

c. seme

d. neme

2.8i **A directional response to a chemical gradient in the environment is known as**

    a. chemotropism

    b. chemotrophism

    c. chemotaxis

    d. chemokinesis

2.9i **The laying of eggs in the nest of another individual of the same species is known as**

    a. interspecific brood parasitism

    b. intraspecific brood parasitism

    c. interspecific brood mutualism

    d. intraspecific brood mutualism

2.10i **Many lizards possess brightly coloured tails that detach when seized by a predator. This is an escape strategy made possible by the phenomenon known as**

    a. autonomy

    b. autolysis

    c. autotomy

    d. autology

2.11i **Which of the following can be divided into avenue builders and maypole builders?**

    a. Rollers

    b. Weaver birds

    c. Sunbirds

    d. Bowerbirds

2.12i **In the common limpet (*Patella vulgata*) reproduction is achieved by several females releasing eggs and several males releasing sperm into the water at the same time (Fig. 2.5). This behaviour is known as**

    a. mass spawning

    b. broadcast spawning

c. scatter spawning

d. dispersal spawning

Fig. 2.5.

2.13i **Spotted hyenas (*Crocuta crocuta*) frequently take the prey killed by African painted dogs (*Lycaon pictus*) in a behaviour known as**

a. kleptoparasitism

b. commensalism

c. obligate parasitism

d. mutualism

2.14i **A sign stimulus is also called a**

a. promoter

b. inducer

c. liberator

d. releaser

2.15i **Birds are able to fly only when they reach a particular age as the result of**

a. learning

b. neuromuscular maturation

c. imprinting

d. conditioning

2.16i Which of the following behaviours has been described as the glue that holds baboon and some other primate societies together?

a. Feeding

b. Mating

c. Grooming

d. Parental care

2.17i The method of kin discrimination whereby an individual's behaviour towards another is determined by how similar they are in appearance, odour or some other observable trait is known as

a. genetic favouritism

b. genomic harmonising

c. phenotypic matching

d. genotypic matching

2.18i Some female birds beg food from their mates during courtship using a behaviour normally only observed in juveniles. As a result of evolution the food begging behaviour has become

a. stereotyped

b. aggrandised

c. stylised

d. ritualised

2.19i The apparatus in Fig. 2.6 consists of a black-and-white chequered raised platform (A) and an identical patterned area located below it (B) that is visible from A through a sheet of glass. This apparatus is used to test depth perception and is called a

a. perception precipice

b. cognition overhang

c. visual canyon

d. visual cliff

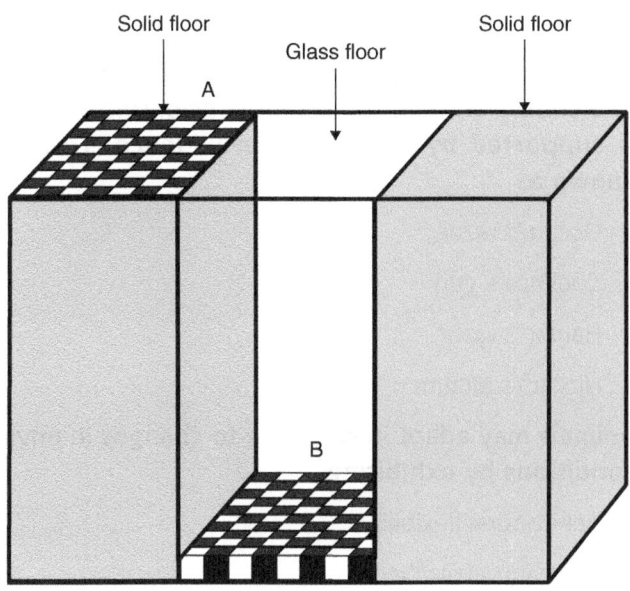

Fig. 2.6.

2.20i **Which of the following statements relating to Fig. 2.6 are true?**

   i.   Chicks of the ground-nesting ring-necked pheasant (*Phasianus colchicus*) almost always avoid stepping on the glass if placed on platform A

   ii. Six-month old human babies will not crawl from platform A onto the glass

   iii. Mandarin duck (*Aix galericulata*) ducklings, which normally hatch in a nest hole in a tree, will not walk from platform A to the edge of the glass

   a. i and ii

   b. i and iii

   c. ii and iii

   d. i, ii and iii

Advanced

2.1a **In attempting to establish the cause of a particular behaviour one should choose the simplest scientific explanation that is supported by the available evidence. This principle is known as**

    a. Occam's razor

    b. Chekhov's gun

    c. Hanlon's razor

    d. Hickam's dictum

2.2a **Animals may adapt in response to changes in environmental conditions by exhibiting**

    a. behavioural flexibility

    b. behavioural plasticity

    c. behavioural pliability

    d. Behavioural elasticity

2.3a **Some lemur species are active periodically and intermittently throughout the day and night. Such animals are referred to as**

    a. matutinal

    b. crepuscular

    c. cathemeral

    d. vespertine

2.4a **A behaviour that results from a conflict between two competing motivations and is irrelevant to the behavioural context is called a**

    a. distraction behaviour

    b. dislocation behaviour

    c. disruption behaviour

    d. displacement behaviour

2.5a **The unique sensory world of an organism is its**

    a. umwelt

    b. ganze

    c. umgebung

    d. verhalten

2.6a ***The Selfish Gene* is a book based around the contention that the more two individuals are genetically related the more sense it makes for them to cooperate with each other. This book was published in 1976 by**

    a. Nancy Moran

    b. Richard Dawkins

    c. Stephen Jay Gould

    d. Jared Diamond

2.7a **Herring gulls (*Larus argentatus*) have a red spot on the tip of the lower bill (Fig. 2.7). Newly-hatched chicks will accurately peck at this spot to induce the adult to regurgitate food. This pecking behaviour is innate and is a**

    a. reaction chain

    b. sign stimulus

    c. fixed action pattern

    d. releaser

Fig. 2.7.

2.8a The two Mauthner neurons in the nervous system of the goldfish (*Carassius auratus*) have evolved to process information about

a. courtship

b. predator attacks

c. feeding

d. buoyancy

2.9a Some bird species engage in 'cooperative breeding' whereby several breeding males and several breeding females use a single communal nest. Females remove the eggs of others in an attempt to maximise their own fitness. This destruction of eggs is called

a. ovicide

b. siblicide

c. patricide

d. matricide

2.10a Many insect species exhibit diapause behaviour. Diapause is a deep resting state they can enter as an adaptation to food shortages in unfavourable conditions. Insects can enter diapause as

i. eggs

ii. mature larvae

iii. pupae

iv. adults

a. iv

b. ii and iv

c. i, ii and iii

d. i, ii, iii and iv

2.11a A behavioural rhythm that occurs over a period of less than 24 hours is

a. a circadian rhythm

b. an infradian rhythm

c. a circannual rhythm

d. an ultradian rhythm

2.12a In amphibians a lingual flip is

a. an acrobatic movement made by some male toads during courtship

b. a method of feeding in frogs and toads whereby the tongue is catapulted out to catch the prey

c. a method of locomotion in salamanders

d. a mating ritual performed on the soil surface by caecilians

2.13a Many animals use materials found in the environment to construct places to live. Which of the following organisms is capable of constructing a 'sand grain house' that it uses as a portable home?

a. The annelid *Eisenia fetida*

b. The lugworm *Arenicola marina*

c. The protozoan *Difflugia coronata*

d. The flagellate *Euglena gracilis*

2.14a Which of the following is the only taxon known to engage in dermatotrophy?

a. Caecilians

b. Salamanders

c. Ragworms

d. Amphioxuses

2.15a A fixed action pattern is a behavioural sequence that is species-specific, highly stereotyped and

a. instinctive

b. learned

c. imprinted

d. conditioned

2.16a Horned lizards such as *Phrynosoma cornutum* and *P. solare* defend themselves by squirting

a. urine from their cloacae

b. blood from blood vessels around their eyes

c. poison from glands in their feet

d. saliva from their mouths

2.17a The degree to which a behaviour can resist being compressed in the time budget of an animal after it has been forced to spend more time on another behaviour, for example foraging, is its

a. behavioural resistance

b. behavioural resoluteness

c. behavioural determination

d. behavioural resilience

2.18a In the fruit fly *Drosophila* the same gene influences the daily activity of the fly and the courtship 'song' of the males that is produced by wing vibrations. The capacity of one gene to have more than one effect on an individual's development is known as

a. polyploidy

b. polygeny

c. pleiotropy

d. phylogeny

2.19a Which of the following made a major contribution to the study of the genetic basis of altruism?

a. William D. Hamilton

b. Arthur Cain

c. Theodosius Dobzhansky

d. Alfred Russel Wallace

2.20a **The contribution that an individual makes to the next generation by supporting related individuals in rearing their young is its**

a. indirect fitness

b. altruistic fitness

c. stabilising selection

d. direct fitness

3 Biological Basis of Behaviour

This chapter contains questions about the structure of the nervous system, the nature of nerve impulses, sense organs and perception, reflexes and the role of hormones in behaviour.

Foundation

3.1f Figure 3.1 is a simplified representation of the structure of a neuron. Match the labels in Table 3.1 with the structures in Fig. 3.1.

Table 3.1

Label	A	B	C	D
R	Schwann cell	Cell body	Schwann cell	Cell body
S	Cell body	Schwann cell	Cell body	Schwann cell
T	Synaptic terminal	Synaptic terminal	Dendrites	Dendrites
U	Dendrites	Dendrites	Synaptic terminal	Synaptic terminal

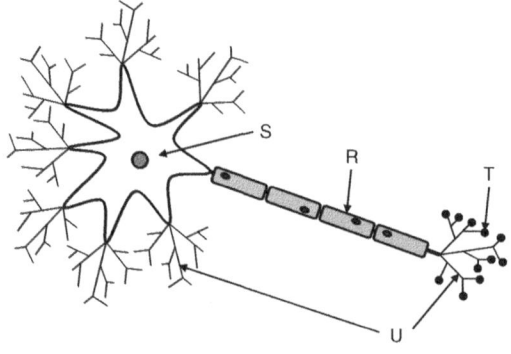

Fig. 3.1.

© Paul A. Rees 2022. *Key Questions in Animal Behaviour and Welfare: A Study and Revision Guide* (P.A. Rees)
DOI: 10.1079/9781789248975.0003

a. A

b. B

c. C

d. D

3.2f The space between the two successive neurons is known as the

a. synaptic fissure

b. neural cleft

c. ganglion

d. synaptic cleft

3.3f The brain and spinal cord comprise the

a. peripheral nervous system

b. central nervous system

c. autonomic nervous system

d. sympathetic nervous system

3.4f A nerve impulse is the result of

a. a reversal of the electrical charge across the membrane of a resting neuron

b. an increase in the quantity of static electricity passing along a neuron

c. a decrease in the quantity of static electricity passing along a neuron

d. a doubling of the electrical potential across the membrane of a neuron

3.5f The vertebrate brain is protected from physical shock by

a. spinal fluid

b. cerebral fluid

c. cerebrospinal fluid

d. lymph

3.6f   In the vertebrate ear the malleus, incus and stapes are known as the

a. tympanic ossicles

b. vestibular ossicles

c. sensory ossicles

d. auditory ossicles

3.7f   The nervous system consists of a nerve net in animals in which of the following taxa?

a. Mollusca

b. Annelida

c. Cnidaria

d. Porifera

3.8f   A lateral line is a system of sense organs in

a. fishes and amphibians

b. reptiles and amphibians

c. birds and reptiles

d. molluscs and annelids

3.9f   The structure that vibrates in the ear of a mammal when it detects a sound is the

a. ventricle

b. septum

c. cochlea

d. tympanum

3.10f   In invertebrates such as arthropods and annelids nerve cords are

a. dorsal and hollow

b. ventral and hollow

c. dorsal and solid

d. ventral and solid

3.11f **Which type of radiation can be detected by pit vipers (Crotalinae) using their pit organs?**

a. Ultraviolet

b. Infrared

c. Microwaves

d. Electromagnetic

3.12f **The medulla oblongata in the vertebrate brain is responsible for**

a. memory

b. homeostatic functions

c. speech

d. vision

3.13f **An effector cell is a**

a. muscle cell

b. gland cell

c. muscle cell or gland cell

d. nerve cell or muscle cell

3.14f **The cerebellum is part of the vertebrate**

a. forebrain

b. midbrain

c. cerebral cortex

d. hindbrain

3.15f **Stretch receptors occur in**

a. the muscles

b. the brain

c. the spinal cord

d. motor neurons

3.16f **The myelin sheath of a neuron is made of**

a. fat

b. protein

    c. protein and fat

    d. carbohydrate and protein

3.17f Which part of the vertebrate brain is involved in vision?

    a. Occipital lobe

    b. Temporal lobe

    c. Parietal lobe

    d. Frontal lobe

3.18f Figure 3.2 is a stylised representation of a reflex arc. Which of the columns in Table 3.2 correctly names the neurons labelled Q, R and S?

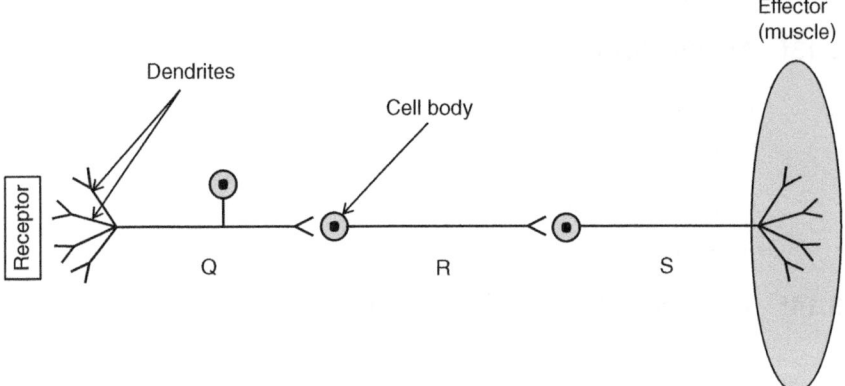

Fig. 3.2.

Table 3.2

Neuron	A	B	C	D
Q	Motor neuron	Sensory neuron	Sensory neuron	Interneuron
R	Interneuron	Interneuron	Motor neuron	Sensory neuron
S	Sensory neuron	Motor neuron	Interneuron	Motor neuron

    a. A

    b. B

    c. C

    d. D

3.19f Ommatidia are components of the

    a. ears of some fishes

    b. eyes of some arthropods

    c. brains of some birds

    d. eyes of some cephalopods

3.20f The two hemispheres of the brain of a placental mammal are connected by a bundle of nerve fibres called the

    a. hypothalamus

    b. central sulcus

    c. pons

    d. corpus callosum

Intermediate

3.1i Which of the diagrams of a section through the axon of a neuron in Fig. 3.3 represents the distribution of electrical charges in an unstimulated neuron (i.e. a neuron in its resting state)?

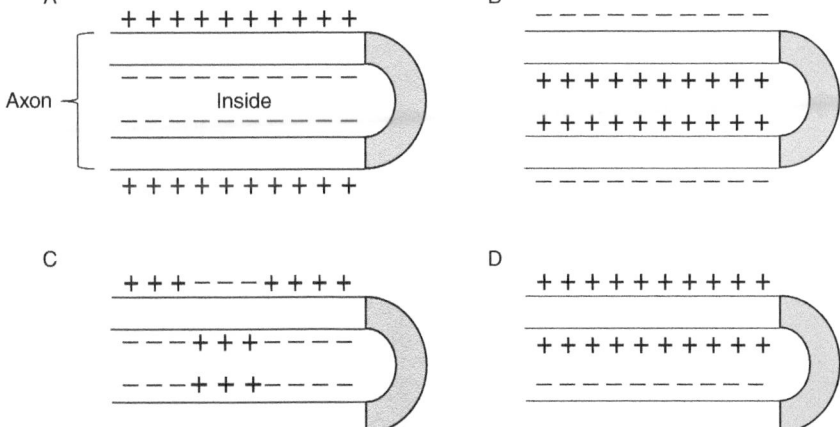

Fig. 3.3.

    a. A

    b. B

c. C

d. D

3.2i   **A nerve impulse travels along the cell membrane of a neuron as a wave of**

a. bipolarisation

b. polarisation

c. depolarisation

d. repolarisation

3.3i   **Figure 3.4 is a diagram of a synapse. Match the labels to the correct structures in Table 3.3.**

Table 3.3

Label	A	B	C	D
P	Synaptic vesicles	Synaptic vesicles	Receptors	Synaptic vesicles
Q	Postsynaptic membrane	Presynaptic membrane	Presynaptic membrane	Presynaptic membrane
R	Receptors	Synaptic cleft	Synaptic vesicles	Receptors
S	Synaptic cleft	Receptors	Synaptic cleft	Synaptic cleft
T	Presynaptic membrane	Postsynaptic membrane	Postsynaptic membrane	Postsynaptic membrane

a. A

b. B

c. C

d. D

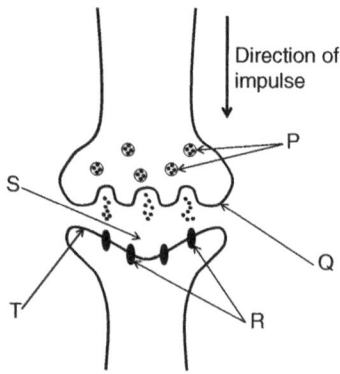

Fig. 3.4.

3.4i **The Jacobson's organ or vomeronasal organ found in amphibians, reptiles and mammals detects**

    a. chemicals

    b. light

    c. sound

    d. vibrations

3.5i **Humboldt squids (*Dosidicus gigas*) are able to change the colour of their skin to produce camouflage, and possibly to communicate with conspecifics, using pigment-filled cells called**

    a. glial cells

    b. erythrocytes

    c. fibroblasts

    d. chromatophores

3.6i **Figure 3.5 shows the changes in membrane potential of an axon when stimulated. Match the labels in the figure with the terms in Table 3.4.**

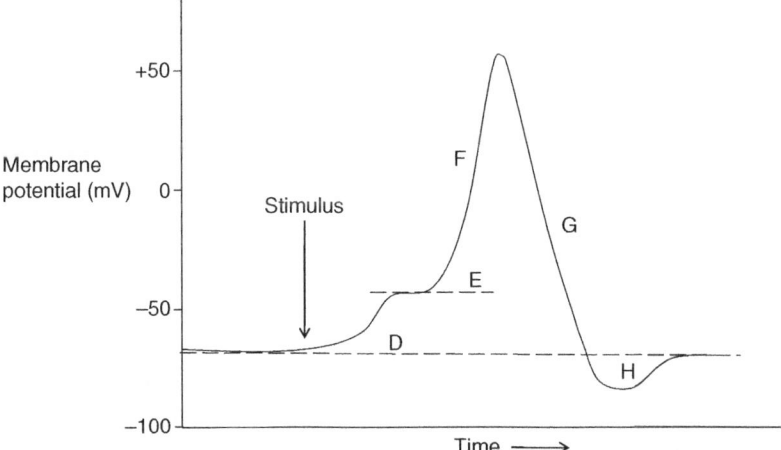

Fig. 3.5.

Table 3.4

	A	B	C	D
D	Resting state	Resting state	Resting state	Threshold potential
E	Refractory period	Threshold potential	Threshold potential	Repolarisation
F	Depolarisation	Depolarisation	Repolarisation	Depolarisation
G	Repolarisation	Repolarisation	Depolarisation	Refractory period
H	Threshold potential	Refractory period	Refractory period	Resting state

    a. A

    b. B

    c. C

    d. D

3.7i **During the 'fight or flight' response which of the following hormones are released?**

    i.   Luteinising hormone

    ii. Adrenaline

    iii. Prolactin

    iv. Noradrenaline

    v. Progesterone

    a. i, ii and iii

    b. ii, iii and v

    c. ii and iv

    d. iii and v

3.8i **Which of the patterns of arrows in Fig. 3.6 (A–D) most accurately illustrates the manner in which a nerve impulse passes along a myelinated neuron? (Note: Arrow X is irrelevant here and relates to Q 3.9i).**

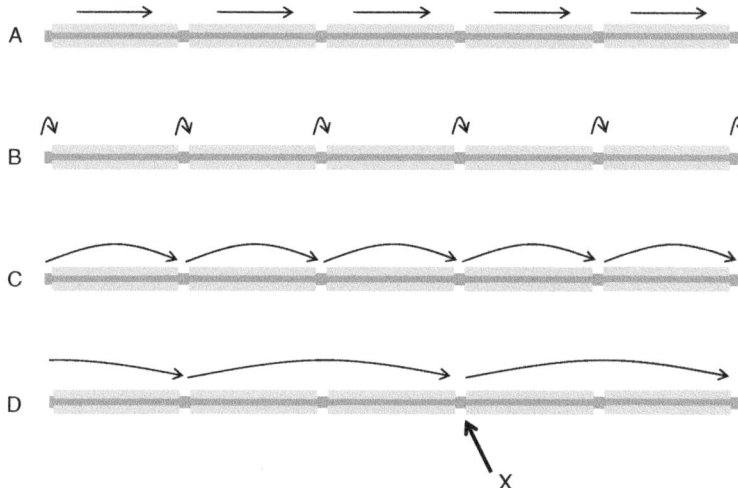

Fig. 3.6.

    a. A

    b. B

    c. C

    d. D

3.9i **In Fig. 3.6 what is the structure labelled 'X'?**

    a. Node of Cuvier

    b. Node of Ranvier

    c. Node of Galen

    d. Node of Curie

3.10i **Proprioreceptors are sense organs that detect**

    a. chemicals

    b. sound

    c. position and movement

    d. light

3.11i **Members of which of the following taxa have a nervous system consisting of radial nerves connected to a nerve ring?**

a. Echinoderms

b. Arthropods

c. Molluscs

d. Annelids

3.12i Cortisol is secreted by the

a. cerebral cortex

b. medulla oblongata

c. adrenal cortex

d. adrenal medulla

3.13i Members of which of the following taxa are considered to possess brains that are as complex as those of some vertebrates?

a. Crustacea

b. Annelida

c. Cephalopoda

d. Echinodermata

3.14i After an impulse has passed along a neuron there is a short period of time during which the neuron cannot be stimulated a second time. This is known as the

a. refractory period

b. resting period

c. reflex period

d. relaxation period

3.15i The region of the vertebrate brain responsible for the unconscious coordination of movement and balance is the

a. cerebral cortex

b. cerebellum

c. cerebrum

d. pons

3.16i In a neuron, neurotransmitters are contained within the

    a. synaptic vesicles

    b. synaptic vacuoles

    c. contractile vacuoles

    d. extracellular vesicles

3.17i Much of the early research on membrane potentials and the nature of nerve impulses was conducted using the axons of

    a. cockroaches

    b. crabs

    c. giant squids

    d. mice

3.18i The section of an axon labelled 'X' in Fig. 3.7 is said to be

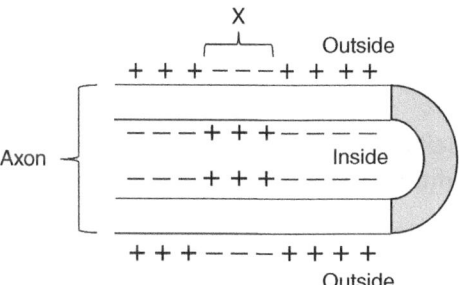

Fig. 3.7.

    a. polarised

    b. depolarised

    c. repolarised

    d. unpolarised

3.19i The mass of adipose tissue that acts as a 'sound lens' in the forehead of dolphins and other toothed whales is called a

    a. melon

    b. pineapple

c. grapefruit

d. mango

3.20i In the vertebrate inner ear the utricle detects

a. low frequency sounds

b. changes in air pressure

c. high frequency sounds

d. linear acceleration and head tilts in the horizontal plane

Advanced

3.1a In deer (Cervidae) the growth, calcification, cleaning (shedding), casting and regeneration of antlers in males are regulated by changes in the secretion of

a. progesterone

b. somatotropin

c. testosterone

d. thyrotropin

3.2a The capacity of nerve cells and neural networks to 'ignore' some stimuli is known as

a. stimulus deletion

b. stimulus avoidance

c. stimulus selection

d. stimulus filtering

3.3a The central stress response system in vertebrates is known as the

a. TBA axis

b. HPA axis

c. HCA axis

d. BPA axis

3.4a   Which of the following enzymes breaks down a neurotransmitter at neuromuscular junctions?

a. Beta-secretase

b. Acetylcholinesterase

c. Transglutaminase

d. RNA methylase

3.5a   Norepinephrine (noradrenaline) functions as

i.   a hormone

ii.   a neurotransmitter

iii. an enzyme

iv.   a visual pigment

a. i and ii

b. iii and iv

c. i and iii

d. ii and iv

3.6a   Maintenance of the normal electrical function of the neuron depends upon the movement of which of the following ions?

a. Na^+ and Cl^-

b. Na^+ and K^+

c. K^+ and Cl^-

d. Na^+ and Fe^-

3.7a   In mammals, the organ of Corti is concerned with the detection of

a. light

b. gravity

c. sound

d. odours

3.8a Endorphins function as

    a. analgesics

    b. vasodilators

    c. diuretics

    d. antiandrogens

3.9a Which of the following parts of the brain is considered to be a link between the nervous system and the endocrine system?

    a. Corpus callosum

    b. Central sulcus

    c. Angular gyrus

    d. Hypothalamus

3.10a The speed at which an impulse passes along a nerve fibre is

    i. directly proportional to the diameter of the fibre

    ii. inversely proportional to the diameter of the fibre

    iii. unaffected by the diameter of the fibre

    iv. higher in myelinated fibres than unmyelinated fibres

    v. slower in myelinated fibres than unmyelinated fibres

    a. i and v

    b. i and iv

    c. ii and v

    d. iii and iv

3.11a A neuron in its resting state has a membrane potential of approximately

    a. −30mV

    b. +30mV

    c. −70mV

    d. +70mV

3.12a **The semicircular canals in the ear of a mammal detect which of the following?**

i. Head rotations caused by self-induced movement

ii. The direction of the force of gravity

iii. Angular accelerations of the head imparted by external forces

iv. Low sound frequencies

a. i

b. ii and iii

c. i and iii

d. ii and iv

3.13a **In invertebrates, statocysts have a role in**

i. balance

ii. vision

iii. hearing

iv. taste

a. i and ii

b. i and iii

c. ii and iii

d. iv

3.14a **When a locust prepares to jump across a gap it moves its head from side to side. This allows it to assess the distance across the gap using a phenomenon known as**

a. retinal displacement

b. disinhibition

c. retinal disparity

d. motion parallax

3.15a Match the names of the types of chromatophores listed in Table 3.5 with the colours they produce.

Table 3.5

Type of chromatophore	A	B	C	D
Melanophore	Yellow	Red	Black/brown	Iridescent white
Xanthophore	Red	Iridescent colours	Yellow	Black/brown
Erythrophore	Iridescent colours	Iridescent white	Red	Yellow
Iridophore	Iridescent white	Black/brown	Iridescent colours	Red
Leucophore	Black/brown	Yellow	Iridescent white	Iridescent colours

    a. A

    b. B

    c. C

    d. D

3.16a The eye of the primitive mollusc *Nautilus* works like a

    a. pinhole camera

    b. digital camera

    c. night vision camera

    d. twin lens reflex camera

3.17a If an insect's halteres are removed it will be unable to

    a. communicate with conspecifics

    b. coordinate its flight

    c. produce visual images

    d. feed

3.18a **Which of the following hormones affects the behaviours associated with parental care in mammals?**

a. prolactin

b. cortisol

c. vasopressin

d. calcitonin

3.19a **Which of the following hormones has/have been shown to have an important function in pair bonding in prairie voles (*Microtus ochrogaster*)?**

a. Oxytocin

b. Vasopressin

c. Oxytocin and vasopressin

d. Glucagon

3.20a **In the mammalian eye, what is the name of the point on the retina behind which the end of the optic nerve is located?**

a. Fovea

b. Vortex ampulla

c. Sclera

d. Blind spot

4 Learning, Memory and Training

This chapter contains questions about the different types of learning exhibited by animals, the relationship between learning and memory, and the principles of animal training.

Foundation

4.1f The modification of voluntary behaviour by the use of a reward is called

   a. imprinting

   b. operant conditioning

   c. reinforcement

   d. classical conditioning

4.2f The German zoologist Richard Semon called a memory trace – the characteristic material trace a stimulus leaves in the nervous system –

   a. a meme

   b. an engram

   c. a deme

   d. a mime

4.3f Teaching a captive rhinoceros to position itself next to a fence for veterinary examination by rewarding it each time it moves near to a circular red board on the fence is known as

© Paul A. Rees 2022. *Key Questions in Animal Behaviour and Welfare: A Study and Revision Guide* (P.A. Rees)
DOI: 10.1079/9781789248975.0004

a. target training

b. fence training

c. obedience training

d. object training

4.4f   In animal training, which of the following could be used as an event marker?

a. A clicker

b. A whistle

c. Verbal praise

d. Any of the above

4.5f   Irene Pepperberg studied the cognitive abilities of *Alex* over a period of 30 years. He could identify and say the names of a wide range of objects, and combine words and phrases together. *Alex* was

a. an Indian rose-ringed parakeet (*Psittacula krameri*)

b. a budgerigar (*Melopsittacus undulatus*)

c. an African grey parrot (*Psittacus erithacus*)

d. a common hill mynah (*Gracula religiosa*)

4.6f   Some noxious species share the same warning signals – for example a black and yellow striped body in some insects – so when predators learn to avoid these signals all noxious species with similar appearance are protected. This phenomenon is called

a. Batesian mimicry

b. Müllerian mimicry

c. Darwinian mimicry

d. Automimicry

4.7f   The strength of the relationship between a stimulus and a consequence depends in part on the nature of the consequence. This principle is known as

a. stimulus relevance

b. stimulus importance

c. stimulus significance

d. stimulus implication

4.8f **Complete the following sentence from the list of terms below: 'Pavlov described the phenomenon whereby a stimulus may be .... so that similar stimuli produce the same response'.**

a. ritualised

b. generalised

c. adapted

d. formalised

4.9f **Most of our knowledge of imprinting comes from**

a. newborn ungulates

b. newborn rodents

c. newly-hatched songbirds

d. newly-hatched galliform and anseriform birds

4.10f **Which part of the mammalian brain is not involved in memory?**

a. Cerebellum

b. Amygdala

c. Pons

d. Hippocampus

4.11f **Captive-bred animals may need to be trained before release to the wild. This training is likely to involve learning to**

a. find and select appropriate food

b. avoid predators

c. react to alarm signals made by conspecifics

d. do all of the above

4.12f As part of the Wildfowl and Wetland Trust's captive breeding programme for the common or Eurasian crane (*Grus grus*) staff developed outfits so that they could disguise themselves as cranes while interacting with the birds (Fig. 4.1). The main purpose of this was to prevent young cranes from

a. developing a fear of humans

b. becoming imprinted on humans

c. exhibiting aggressive behaviour towards staff

d. escaping

Fig. 4.1.

4.13f Figure 4.2 illustrates an example of

a. operant conditioning

b. instrumental conditioning

c. classical conditioning

d. Type II conditioning

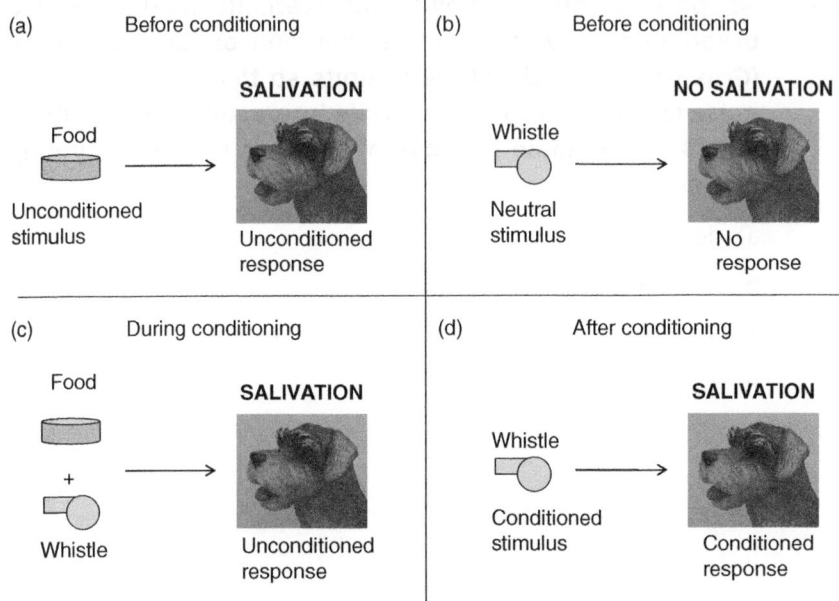

Fig. 4.2.

4.14f A mouse is taught to run a complex maze for a food reward. Which of the following statements about this experiment is likely to be false?

a. Errors will increase with the number of trials

b. Time taken to run the maze will decrease with the number of trials

c. A hungry mouse is likely to perform better than a well-fed mouse

d. The mouse must always begin the maze in the same place

4.15f What was defined by Thorpe as 'the association of indifferent stimuli or associations without patent reward'?

a. Insight learning

b. Latent learning

c. Reasoning

d. Habituation

4.16f **The learning process in which an animal's reaction to an event depends upon the fact that it has encountered an event only once previously is known as**

a. associative learning

b. imprinting

c. single exposure conditioning

d. one event learning

4.17f **Animals living on farms and in zoos are often retained behind electric fences (hot wires). The fences themselves are not strong enough to stop the animals from escaping from their fields or enclosures but they learn that contact with the fence results in a mild electric shock. The type of learning involved is**

a. classical conditioning

b. habituation

c. trial-and-error learning

d. imprinting

4.18f **Which of the following first showed that some animals are able to learn by a 'sudden insight' into the 'structure' of a problem?**

a. Wolfgang Köhler

b. Konrad Lorenz

c. Frans de Vaal

d. Harry Harlow

4.19f **A group of bees were regularly fed from a dish placed on a square piece of blue paper. When presented with two squares of paper, one red and one blue, neither of which contained food, they settled on the blue paper and ignored the red paper. The same bees were exposed to a table containing one blue square and three grey squares, each of which was a different shade of grey; all four squares contained an empty food dish. All of the bees settled on the blue square. The experiment demonstrated that the bees were**

a. habituated to the colour blue and have monochrome vision

b. conditioned to the colour blue and have monochrome vision

c. habituated to the colour blue and have colour vision

d. conditioned to the colour blue and have colour vision

4.20f When an animal is being trained it is often difficult to reward it sufficiently close to the time at which it performs the desired behaviour to ensure that an association is made between that behaviour and the reward. In order to link the two the trainer will use a 'device' as soon as the desired behaviour is performed known as a

a. tie

b. bridge

c. linkage

d. bond

Intermediate

4.1i Trial-and-error learning is also called

a. operant conditioning

b. classical conditioning

c. habituation

d. Pavlovian conditioning

4.2i Figure 4.3(a) shows the time taken for a small rodent to complete a maze (i.e. travel from X to food at Y)(Fig. 4.3(b)) in successive trials. The graph shows

a. a memory curve

b. a learning curve

c. a success curve

d. an achievement curve

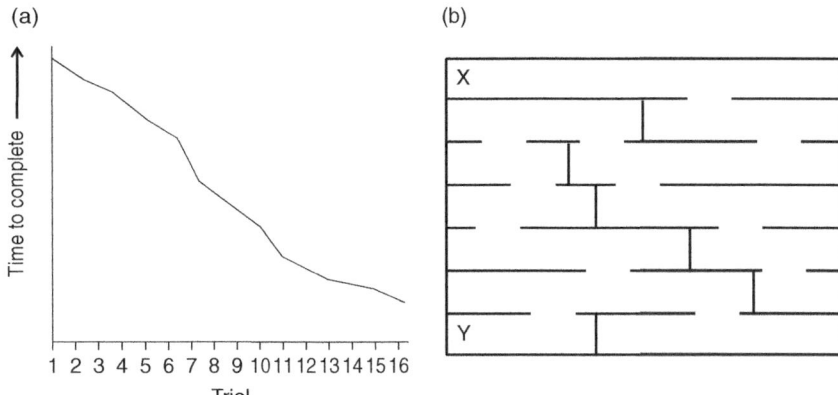

Fig. 4.3.

4.3i **Which of the following types of memory is concerned with remembering the general knowledge accumulated throughout life?**

a. Implicit memory

b. Nondeclarative memory

c. Semantic memory

d. Procedural memory

4.4i **The transfer of information from short-term memory to long-term memory is enhanced by**

a. rehearsal

b. positive or negative emotional states mediated by the amygdala

c. the association of new data with data already previously learned and stored in long-term memory

d. all of the above

4.5i **When negative reinforcement is used in animal training it involves**

a. the removal of an unpleasant stimulus

b. punishment

c. the addition of a pleasant stimulus

d. negative punishment

4.6i **Which of the following statements about imprinting is false?**

a. It has been widely reported from studies of birds

b. It is a type of learning

c. Sexual imprinting is important in the normal development of many young animals

d. It does not have to occur during a critical period after birth or hatching

4.7i **Which of the following statements is true about the scientists who identified different types of mimicry in animals?**

a. Fritz Müller was an Austrian zoologist and Henry Bates was an American naturalist

b. Fritz Müller was a German zoologist and Henry Bates was an English naturalist

c. Fritz Müller was a Swiss zoologist and Henry Bates was an English naturalist

d. Fritz Müller was an American zoologist and Henry Bates was a Scottish naturalist

4.8i **The Asian elephant (*Elephas maximus*) in Fig. 4.4 placed a log on the ground under the suspended net and then stood on it with her front feet so that she could reach the food. This is an example of**

a. forethought learning

b. foresight learning

c. insight learning

d. hindsight learning

Fig. 4.4.

4.9i   During animal training primary rewards and second-order reinforcers may be used. In Table 4.1 which column accurately lists the four objects in the correct category?

Table 4.1

Category	A	B	C	D
Primary rewards	Treat	Clicker	Treat	Toy
	Clicker	Whistle	Toy	Whistle
Second-order reinforcers	Whistle	Toy	Clicker	Treat
	Toy	Treat	Whistle	Clicker

a. A

b. B

c. C

d. D

4.10i **In the process of teaching a puppy to sit the responses of the trainer to the various intermediate behaviours were as indicated in Table 4.2.**

Table 4.2

Stage	Behaviour of puppy	Trainer's response
1	Slight squat	Reward given
2	Squat resembling a sit	Reward given
	Slight squat	No longer rewarded
3	True sit	Rewarded
	Squat resembling a sit	No longer rewarded
	Slight squat	No longer rewarded

This process is called

a. framing

b. shaping

c. moulding

d. forming

4.11i **The process by which a young mammal learns to identify a mate of the same species is known as**

a. sexual selection

b. sexual imprinting

c. sexual dimorphism

d. sexual orientation

4.12i **When a conditioned stimulus is presented repeatedly without the unconditioned stimulus the response may be reduced and eventually disappear. This phenomenon is known as**

a. eradication

b. annihilation

c. extermination

d. extinction

4.13i **Which of the following scientists used puzzle boxes extensively in his research, especially to study trial-and-error learning in cats?**

   a. Edward Thorndike

   b. Konrad Lorenz

   c. Wolfgang Köhler

   d. Harry Harlow

4.14i **Match the letters in Fig. 4.5 with the correct terms in Table 4.3 to describe the process of clicker training.**

Fig. 4.5.

Table 4.3

Label	A	B	C	D
Q	Trainer uses clicker	Trainer uses clicker	Command given by trainer	Command given by trainer
R	Command given by trainer	Command given by trainer	Correct response from animal	Correct response from animal
S	Correct response from animal	Trainer gives reward to animal	Trainer uses clicker	Trainer gives reward to animal
T	Trainer gives reward to animal	Correct response from animal	Trainer gives reward to animal	Trainer uses clicker

   a. A

   b. B

   c. C

   d. D

4.15i **Which of the following may be described as a type of phase-sensitive learning?**

   a. habituation

   b. classical conditioning

   c. insight learning

   d. imprinting

4.16i **The part of memory that remembers specific past experiences based on unique 'what-where-when' components is known as**

   a. temporary memory

   b. episodic memory

   c. semantic memory

   d. specific memory

4.17i **Sensitisation is the opposite of**

   a. habituation

   b. aversion

   c. conditioning

   d. stimulation

4.18i **During training, the continued presentation of a reinforcer may reduce its effectiveness. For example, an animal receives a total of 5kg of meat during a training session instead of the usual 1kg. This may result in the animal refusing to train for several hours due to a phenomenon known as**

   a. fulfilment

   b. repletion

   c. satisfaction

   d. satiation

4.19i **Müllerian mimicry was first identified in**

   a. American beetles

   b. European moths

   c. subtropical ants

   d. tropical butterflies

4.20i **In Pavlov's study of salivation in dogs, the bell was**

   a. the unconditioned reflex

   b. the conditioned reflex

c. the conditioned stimulus

d. the unconditioned stimulus

Advanced

4.1a **Experimental evidence was reported by Bédécarrats _et al._ (2018) suggesting that memories can be transferred from trained snails (_Aplysia californica_) to untrained snails via extracts of the former's**

a. DNA

b. RNA

c. ADP

d. ATP

4.2a **A schedule of reinforcement whereby the reinforcement is delivered after a predictable amount of time is known as**

a. a fixed interval schedule

b. scheduled reinforcement

c. a 1: many ratio schedule

d. a final response ratio

4.3a **A pigeon receives a stimulus (a food reward) after every tenth time it pecks at a button. This type of reward schedule is a**

a. variable interval schedule

b. final response schedule

c. variable ratio schedule

d. fixed ratio schedule

4.4a **Select the correct missing words (K, L, M and N) to complete the descriptions of reinforcement and punishment in Table 4.4 from the options in Table 4.5.**

Table 4.4

	Reinforcement	Punishment
Positive	……..(K) stimulus added to increase a behaviour	……..(L) stimulus added to decrease a behaviour
Negative	……..(M) stimulus removed to increase a behaviour	……..(N) stimulus removed to decrease a behaviour

Table 4.5

	A	B	C	D
K	Unpleasant	Pleasant	Pleasant	Unpleasant
L	Pleasant	Unpleasant	Unpleasant	Pleasant
M	Pleasant	Unpleasant	Pleasant	Unpleasant
N	Unpleasant	Pleasant	Unpleasant	Pleasant

    a. A

    b. B

    c. C

    d. D

4.5a **Animals may be taught complex behaviours by dividing them into segments and then conditioning them individually. This is called**

    a. shaping

    b. chaining

    c. linking

    d. merging

4.6a *Monty*, *Ginny* and *Porter* were three domestic dogs in New Zealand that became famous because they had been taught to

    a. play football

    b. skateboard

    c. drive a car

    d. play a computer game

4.7a The nematode *Caenorhabditis elegans* exhibits a memory-based behaviour related to its cultivation temperature. Well-fed animals cultivated between 15°C and 25°C and placed on a thermal gradient migrate towards their cultivation temperature. When they reach it they move in straight lines along points at this temperature. The behaviour that maintains the animal in its preferred temperature is known as

    a. positive thermotaxis

    b. negative thermotaxis

    c. isothermal tracking

    d. anisothermal tracking

4.8a The socially-learned behaviour of washing sweet potatoes, whereby the behaviour spread to others from a single individual, was famously studied in a troop of

    a. Barbary apes (*Macaca sylvanus*)

    b. Hanuman langurs (*Semnopithecus entellus*)

    c. Golden snub-nosed monkey (*Rhinopithecus roxellana*)

    d. Japanese macaques (*Macaca fuscata*)

4.9a The process by which an animal in a group is induced to perform a behaviour after observing its performance by others – often resulting in individuals co-ordinating their behaviour so that they, for example, feed at the same time – with some degree of mutual stimulation is called

    a. social facilitation

    b. social interaction

    c. social mimicry

    d. socialisation

4.10a Some birds memorise their species' song during a sensitive period and improve it with practice until it sounds like the adult song. Match the elements K, L and M in Fig. 4.6 with the correct terms in Table 4.6.

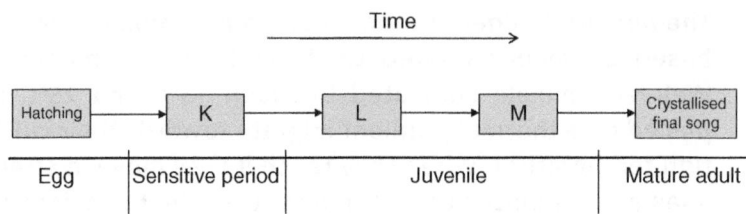

Fig. 4.6.

Table 4.6

Element	A	B	C	D
K	Passive listening	Template	Passive listening	Subsong
L	Subsong	Passive listening	Template	Passive listening
M	Template	Subsong	Subsong	Template

    a. A

    b. B

    c. C

    d. D

4.11a **Which of the following species have been trained to detect trinitrotoluene (TNT), a common component of land mines?**

    i. African elephants (*Loxodonta africana*)

    ii. Domestic dogs (*Canis lupus familiaris*)

    iii. African giant pouched rats (*Cricetomys gambianus*)

    iv. Ratels (*Mellivora capensis*)

    a. ii

    b. ii and iv

    c. i, ii and iii

    d. ii and iii

4.12a The growth architecture of a clover (*Trifolium repens*) branch may be influenced by its current neighbours and also by the neighbours it had interacted with in the previous year (Turkington *et al.*, 1991). This suggests that its growth form is affected by its accumulated experiences. This may be thought of as a type of

a. memory

b. sentience

c. conditioning

d. taxis

4.13a **An animal may reject food that it had ingested on a previous occasion if its ingestion was followed by gastrointestinal upset. This is known as the**

a. Mercia effect

b. Garibaldi effect

c. Medusa effect

d. Garcia effect

4.14a **Tryon's Rat Experiment was a multi-decade selective breeding experiment in which he measured the abilities of successive generations of rats to complete a maze (Tryon, 1940). The purpose of the experiment was to determine whether or not**

a. genetic traits contribute to behaviour

b. the type of food reward influenced performance

c. only certain environmental factors affect behaviour

d. performance was affected by age

4.15a **The time it takes an animal to learn to associate two events depends upon how closely they occur in time. In three experiments with chickens the time a worm made a sound was recorded as zero. The time the chicken found the worm was recorded as the number of seconds after the sound occurred (Table 4.7). Match each of the experiments listed in Table 4.7 with the correct result in Table 4.8 where 'Slowly' means an association between the sound and the presence of a worm was made slowly (after many events) and 'Rapidly' means the association was made quickly (after few events).**

Table 4.7

Experiment	Time of event	
	Sound of worm	Chicken finds worm
K	0 sec	10 sec
L	0 sec	60 sec
M	0 sec	0 sec

Table 4.8

Experiment	A	B	C	D
K	Slowly	Slowly	Rapidly	Slowly
L	Slowly	Rapidly	Slowly	Rapidly
M	Rapidly	Slowly	Slowly	Rapidly

    a. A

    b. B

    c. C

    d. D

4.16a **When the ammonia from cows' urine mixes with soil it results in the release of the greenhouse gas nitrous oxide. Cows have been trained to use a toilet pen (*MooLoo*) so that the urine can be collected and treated. The cows were placed in the *MooLoo* and rewarded with food when they urinated. They were then placed in an area adjacent to the *MooLoo* and rewarded if they walked into it and urinated. Cows who urinated outside of the *MooLoo* were sprayed with water for 3 seconds. Training the cows to enter and use the *MooLoo* was achieved using**

    a. positive reinforcement and positive punishment

    b. positive reinforcement and negative punishment

    c. negative reinforcement and positive punishment

    d. negative reinforcement and negative punishment

4.17a **Social learning in fishes has been demonstrated with respect to**

i.   food choice

ii. food location

iii. mate choice

iv. predator recognition

a. i and ii

b. iii and iv

c. i, ii and iv

d. i, ii, iii and iv

4.18a **Which of the following is the simplest organism that has been shown experimentally to exhibit learning as a result of Pavlovian conditioning?**

a. Goldfish (*Carassius auratus*)

b. Sponge (Porifera)

c. *Paramecium* (a single celled ciliate)

d. Earthworm (*Lumbricus terrestris*)

4.19a. **Which of the following structures is critical in the encoding of the environmental information in map-like or relational memory representations for spatial navigation in mammals, birds and reptiles?**

a. Hippocampus

b. Hypothalamus

c. Pituitary gland

d. Thymus

4.20a **In a study of the role of learning in the feeding behaviour of the fifteen-spined stickleback (*Spinachia spinachia*) Croy and Hughes (1991) found that these fish experienced difficulties in exploiting different feeding strategies for different types of prey when fed alternating rather than pure diets;**

they learned less efficiently and had shorter memories. Retrieval of memories was affected by

a. intervention

b. interference

c. intrusion

d. intercession

5 Territoriality, Navigation and Migration

This chapter contains questions about home range, territoriality, the methods used by animals in orientation and navigation, and the mechanisms that control migration.

Foundation

5.1f   Migration is a special type of

    a. dispersion

    b. dispersal

    c. disposal

    d. disposition

5.2f   The geographical area occupied by an entire species is its

    a. territory

    b. home range

    c. domain

    d. range

5.3f   Animals that are able to head in a particular direction without reference to landmarks using, for example, information from the Earth's magnetic field, are said to possess a

    a. theodolite

    b. GPS

© Paul A. Rees 2022. *Key Questions in Animal Behaviour and Welfare: A Study and Revision Guide* (P.A. Rees)
DOI: 10.1079/9781789248975.0005

c. compass

d. sextant

5.4f   The ability of an animal to return to a place from which it has been displaced is called

a. taxis

b. kinesis

c. ranging

d. homing

5.5f   Individuals of which of the following species mark their territory with cubical faeces?

a. Common wombat (*Vombatus ursinus*)

b. Koala (*Phascolarctos cinereus*)

c. Tasmanian devil (*Sarcophilus harrisii*)

d. Eastern quoll (*Dasyurus viverrinus*)

5.6f   The ability of green turtles (*Chelonia mydas*) to migrate from Brazil and find the Ascension Island (2,300 km away) may be explained from an understanding of

a. ocean currents

b. continental drift

c. lunar cycles

d. tides

5.7f   Adult female green turtles migrate from the coast of Brazil to Ascension Island in the Atlantic Ocean to

a. avoid predators

b. find food

c. lay their eggs

d. find mates

5.8f   In a study published in 1976 researchers reported exposing juvenile fish to morpholine or phenethyl alcohol for 1.5 months and then releasing them in Lake Michigan (Scholz *et al.*, 1976). Eighteen months later, during the spawning

migration, they metered the two chemicals into separate streams and counted the number of morpholine-exposed and phenethyl alcohol-exposed fish returning to each stream. Most of the fish that had been exposed to morpholine were captured in the stream scented with morpholine and most of those exposed to phenethyl alcohol were captured in the stream treated with phenethyl alcohol. The researchers concluded that the fish had imprinted on the chemicals to which they were exposed and used chemical cues in homing. The fish were most likely to have been a species of

a. salmon

b. cichlid

c. eels

d. herring

5.9f **Which of the following taxa have at least some species that are capable of echolocation?**

i. Cetaceans

ii. Bats

iii. Birds

iv. Shrews

v. Tenrecs

a. i and ii

b. i, ii and iv

c. i, ii, iii and v

d. i, ii, iii, iv and v

5.10f **Which of the following statements about territoriality is false?**

a. All territories held by members of the same species are approximately the same size

b. Contests over territories may consist of vocalisations

c. Fights over territories are rarely fatal

d. Fights between holders of adjacent territories are most common at the boundary between the territories

5.11f The difference between a territory and a home range is

a. a territory is defended, a home range is not

b. a home range is defended, a territory is not

c. the territory of a species is always larger than its home range

d. a territory is occupied by a male, a home range is occupied by a female.

5.12f Which of the following could not be the basis of a celestial compass used by migratory birds?

a. The positions of the stars

b. The position of the Sun

c. Skylight polarisation patterns

d. The alignment of the Earth's magnetic field

5.13f Humpback whales (*Megaptera novaeangliae*) migrate between

a. Arctic feeding grounds and Antarctic mating and calving grounds

b. Arctic mating and calving grounds and Antarctic feeding grounds

c. polar feeding grounds and tropical mating and calving grounds

d. polar mating and calving grounds and tropical feeding grounds

5.14f Which of the following is a form of defence of breeding sites, food sources and other resources from conspecifics?

a. Ranging behaviour

b. Resource partitioning

c. Territoriality

d. Intraspecific competition

5.15f Males of some bird species hold more than one territory during the breeding season. This phenomenon is known as

a. polyterritoriality

b. multiterritoriality

c. metaterritoriality

d. megaterritoriality

5.16f **White rhinoceros bulls (*Ceratotherium simum*) mark their territories with piles of dung known as**

a. latrines

b. middens

c. scats

d. coprolites

5.17f **Individuals of which of the following species do not perform a handstand to scent mark vertical surfaces?**

a. Ring-tailed lemur (*Lemur catta*)

b. Giant panda (*Ailuropoda melanoleuca*)

c. Red squirrel (*Sciurus vulgaris*)

d. Dwarf mongoose (*Helogale parvula*)

5.18f **In many migrant bird species the two sexes lead separate lives for most of the year but join up with a partner each breeding season. This behaviour is called**

a. perennial monogamy

b. seasonal monogamy

c. seasonal polygyny

d. annual polygyny

5.19f **Which of the following birds has the longest annual migration?**

a. Northern gannet (*Morus bassanus*)

b. Arctic skua (*Stercorarius parasiticus*)

c. Wandering albatross (*Diomedea exulans*)

d. Arctic tern (*Sterna paradisaea*)

5.20f **Some species, for example the common crossbill (*Loxia curvirostra*) undergo periodic sudden migrations in response to food availability. This is known as**

a. irruptive migration

b. differential migration

c. partial migration

d. disruptive migration

Intermediate

5.1i **In general, which of the following are the potential costs and which are potential benefits of territoriality (Table 5.1)?**

Table 5.1

Cost/benefit	A	B	C	D
Exclusion of intruders	Cost	Cost	Benefit	Benefit
Access to food	Benefit	Benefit	Cost	Cost
Access to sheltering places	Benefit	Benefit	Cost	Benefit
Access to mates	Cost	Benefit	Cost	Benefit
Direct combat	Cost	Cost	Benefit	Cost
Access to breeding sites	Benefit	Benefit	Cost	Benefit

a. A

b. B

c. C

d. D

5.2i **Which of the following statements about migration in monarch butterflies (*Danaus plexippus*) is false?**

a. Monarchs are the only butterfly species known to undertake a two-way migration

b. No single butterfly completes the entire migration

c. The butterflies migrate between California and Florida

d. The butterflies produce several generations each year

5.3i **The tendency of an animal to return to the site of its birth to breed is known as**

a. natal polyandry

b. natal pleiotropy

c. natal polygeny

d. natal philopatry

5.4i **Which of the following species engages in mass migrations in response to population increases?**

a. Greater cane rat (*Thryonomys swinderianus*)

b. Norway lemming (*Lemmus lemmus*)

c. Azara's agouti (*Dasyprocta azarae*)

d. Syrian hamster (*Mesocricetus auratus*)

5.5i **Fish that migrate from seawater to freshwater habitats to spawn are referred to as**

a. anadromous

b. catadromous

c. diadromous

d. endogenous

5.6i **An animal that has an ability to travel in a particular direction without reference to landmarks in conjunction with a biological clock possesses a**

a. homing compass

b. capacity for mental time travel

c. time-compensated compass

d. time-adjusted mental map

5.7i **Sonar used by the United States Navy disrupts the movement patterns of, and causes mass strandings in, some species of**

a. manatees

b. cetaceans

c. sea otters

d. seabirds

5.8i **The structure labelled X in Fig. 5.1 is used by the males of some lizard species as a signalling device to attract females and to indicate territorial boundaries. It is called**

Fig. 5.1.

a. a dewlap

b. a mandible

c. a forelock

d. a beard

5.9i **A single resource territory is so called because it is defended**

a. by a single male

b. for a single purpose

c. for a single breeding season

d. by a single pair

5.10i **Feeding territory size in birds of prey that feed on small mammals is likely to be**

    a. unrelated to small mammal density

    b. positively correlated with small mammal density

    c. negatively correlated with small mammal density

    d. directly proportional to small mammal density

5.11i **Digger wasps are solitary species that make nests underground that are accessible via a small entrance hole. An ethologist marked the nest entrance of a female digger wasp with a ring of pine cones. Several days later, when the wasp was away from the nest, she replaced the circle of pine cones with four different arrangements of objects (Fig. 5.2A–D) approximately 2 metres from the nest entrance. When the wasp returned she was most likely to search for the nest entrance in the centre of which of the arrangements of objects?**

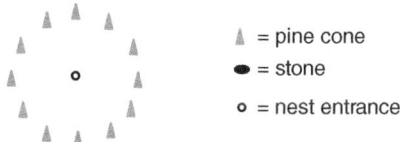

    ▲ = pine cone

    ● = stone

    ○ = nest entrance

A	B	C	D

Fig. 5.2.

    a. A

    b. B

    c. C

    d. D

5.12i **Evidence for the existence of a magnetic compass was found by William Keeton of Cornell University in which of the following animals?**

    a. Monarch butterflies (*Danaus plexippus*)

    b. Homing pigeons (*Columba livia domestica*)

    c. European storks (*Ciconia ciconia*)

    d. Humpback whales (*Megaptera novaeangliae*)

5.13i **Individuals of which of the following species mark their territories with chemicals produced by preorbital scent glands?**

    a. Giraffe (*Giraffa camelopardalis*)

    b. Okapi (*Okapia johnstoni*)

    c. Greater one-horned rhinoceros (*Rhinoceros unicornis*)

    d. Oribi (*Ourebia ourebi*)

5.14i **The response whereby a fish turns to face upstream in a river against the water current is known as**

    a. positive rheotaxis

    b. negative tropotaxis

    c. negative rheotaxis

    d. positive tropotaxis

5.15i **The use of polarised light as a directional cue is particularly useful for animals that do not have a direct view of the**

    a. Moon

    b. stars

    c. horizon

    d. Sun

5.16i **Which of the following is not the name of one of the flyways used by migratory birds flying across North America?**

    a. Pacific flyway

    b. Mississippi flyway

c. Midwest flyway

d. Atlantic flyway

5.17i Birds migrate to balance their

a. protein and carbohydrate intake

b. energy intake and expenditure

c. internal body temperature

d. fluid intake

5.18i Navigation by means of landmarks is called

a. pharotaxis

b. pharotropism

c. geotaxis

d. geotropism

5.19i Polarised light is used by some species in orientation. It is a product of the scattering of light by the atmosphere which causes light waves to vibrate primarily in a plane that is at which of the following angles to the direction of wave propagation?

a. 25°

b. 45°

c. 90°

d. 120°

5.20i A warbler that normally flies north in the spring and south in the autumn may be induced to fly north in the autumn by

a. providing it with additional food

b. reducing its energy intake

c. injecting it with hormones to bring it into breeding condition

d. injecting it with hormones to prevent it coming into breeding condition

Advanced

5.1a Which of the following do birds use to migrate?

a. Celestial compass clues

b. Mental maps of geographical references

c. The Earth's magnetic field

d. All of the above

5.2a In some species individuals exhibit reduced aggression towards familiar conspecific neighbours compared with strangers to reduce the cost of territory defence. This is known as the

a. economic defence effect

b. best friend effect

c. close neighbour effect

d. dear enemy effect

5.3a A feature of an animal's external environment that provides it with information about the passage of time is called a

a. wetter

b. jahreszeiten

c. zeitgeber

d. wechselgeld

5.4a The satellite system used to track the movements of animals such as geese and marine mammals from space is known as

a. Argos

b. Opus

c. Naxos

d. Delos

5.5a The blackpoll warbler (*Setophaga striata*) migrates from eastern Canada to South America across the Atlantic Ocean. Which of the following is unlikely to be a reason

why this route is preferable to that which would take the warblers' journey over land?

a. It is much shorter

b. It is likely to result in fewer encounters with predators

c. Easterly winds in the North Atlantic give the birds a tail wind at the start of their journey

d. Easterly breezes in the South Atlantic help warblers reach islands in the Atlantic and the Caribbean

5.6a **The European blackbird (*Turdus merula*) has migratory and non-migratory types. Which of the following statements is false for this species?**

a. Birds that were resident in one winter rarely switched to being migratory in the next winter

b. Birds that were migratory in one winter often became residents in the next winter

c. Birds that switch from being migratory to resident are usually younger birds

d. The two pure strategies hypothesis is unlikely to apply to this species

5.7a **In the speckled wood butterfly (*Pararge aegeria*) males compete for territories consisting of spots of sunlight on the ground layer of woodland (Davies, 1978). Males fly down from the canopy to occupy sunspots. If a sunspot is occupied,**

a. the intruder always leaves of its own accord

b. there is a high probability an intruder will displace the territory owner

c. the intruder is always driven away by the territory owner even if he has only occupied the sunspot for just a few seconds

d. the intruder will be killed by the territory owner

5.8a In some songbirds it has been observed that males that returned to the same territory had reproduced more successfully in the preceding year than had males that did not return to their old breeding grounds. This behaviour – males returning to sites where they had previously bred successfully – is called

    a. site consistency

    b. site fidelity

    c. site allegiance

    d. site constancy

5.9a Which of the following have a magnetic sense used in orientation?

    a. Lobsters

    b. Pigeons

    c. Frogs

    d. All of the above

5.10a In the dunnock or hedge sparrow (*Prunella modularis*) – a small songbird – a female sometimes lives with two males (the alpha and the beta male) in a territory. If the socially subordinate beta male has copulated with the female during the egg-laying period he may provide parental care for her brood even though he may not be the father. Alpha and beta duos adjust the amount of parental effort they expend depending upon

    a. the availability of food

    b. the exclusivity of their sexual access to the female

    c. the size of the territory

    d. the total number of chicks

5.11a Krebs (1982) removed great tits (*Parus major*) from their territories and then measured the duration of fights with the new residents when the original territory holders were returned (Fig. 5.3). Which of the statements below is not supported by the data used to construct Fig. 5.4?

Fig. 5.3.

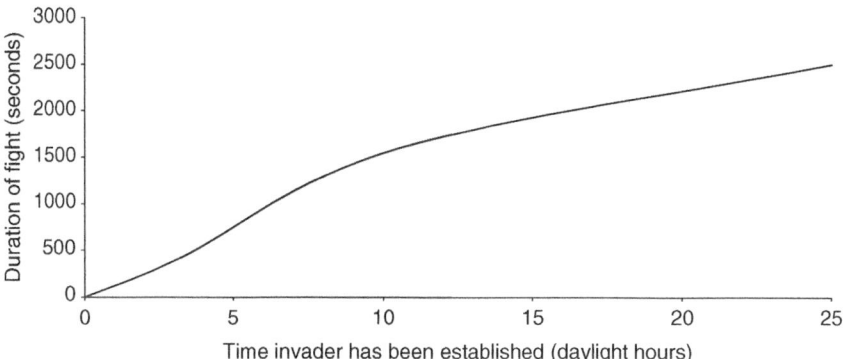

Fig. 5.4.

a. After a tenure of 25 hours the fight duration was four times higher than after a tenure of 5 minutes

b. The fight duration increased at a faster rate in the first ten-minute period of tenure than in the second ten-minute period of tenure

c. There is a positive correlation between duration of the invader's tenure and the duration of the fight

d. The original territory owner always won the fight

5.12a The most important factors controlling the annual migration of wildebeest (*Connochaetes taurinus*) between the Serengeti in Tanzania and the Masai Mara in Kenya (Fig. 5.5) are

i. rainfall

ii. predation pressure

iii. temperature changes

iv. plant nutritional gradients

a. i and ii

b. i and iii

c. ii and iv

d. i and iv

Fig. 5.5.

5.13a **The concept of home range was first defined by Burt (1943) in relation to**

a. birds

b. insects

c. mammals

d. reptiles

5.14a **'Time-plan spacing', in which males use and scent mark the same area at different times – similar to the manner in which different families use the same holiday apartment in different weeks of the year on a time-share basis – is used by which of the following species?**

i. Asian elephant (*Elephas maximus*)

ii. Cheetah (*Acinonyx jubatus*)

iii. Domestic cat (*Felis catus*)

iv. Hippopotamus (*Hippopotamus amphibius*)

a. i only

b. ii and iii

c. i, ii and iii

d. ii, iii and iv

5.15a **When moving from the beach where they have hatched to the open ocean loggerhead turtle (*Caretta carreta*) hatchlings use cues in which of the following sequences (starting on the beach)?**

a. Magnetic orientation → visual cues → wave orientation → wave direction

b. Visual cues → magnetic orientation → wave direction → wave orientation

c. Visual cues → wave orientation → magnetic orientation → wave direction

d. Visual cues → magnetic orientation → wave orientation → wave direction

5.16a **The home range of an animal may be estimated by obtaining data on its location at various points in time – for example from GPS equipment – and producing a map of these locations. Fig. 5.6 shows one method of delineating a home range from these data. This method is called the**

a. maximum convex polygon method

b. minimum convex polygon method

c. minimum concave polygon method

d. optimal polygon method

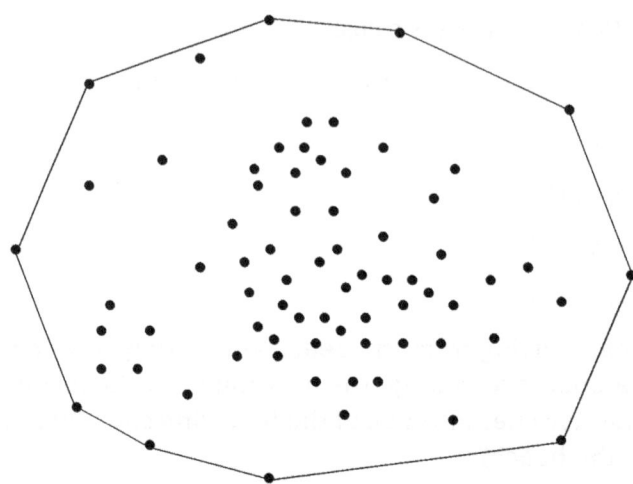

Fig. 5.6.

5.17a ***The Evolutionary Ecology of Animal Migration*** **is a major textbook written by**

   a. George Schaller

   b. Richard Estes

   c. Edward O. Wilson

   d. Robin Baker

5.18a **Briefer *et al.* (2008) performed an experiment with skylarks (*Alauda arvensis*) in which they played the songs of their neighbours and those of strangers at different times in the breeding season. When during the breeding season did they find that neighbours were treated as 'dear enemies'?**

   a. Throughout

   b. At the beginning

   c. In the middle

   d. At the end

5.19a **Two groups of pigeons were used in an experimental study of bird navigation. Group 1 had bar magnets attached to their backs. Group 2 had brass bars of the same mass as the magnets attached to their backs. All of the pigeons**

were released 20 miles from their home loft. Which of the following statements is false?

a. The pigeons carrying the brass bars were the control group

b. In sunny conditions all birds from both groups returned home quickly and safely

c. In overcast conditions the birds in Group 2 were disorientated

d. In overcast conditions the birds in Group 1 were disorientated

5.20a Which of the following hormones play a part in migration in various vertebrate species?

i. Prolactin

ii. Thyroid hormones

iii. Cortical steroids

iv. Gonadotrophins

v. Gonadal steroids

a. i, ii and iv

b. iii, iv and v

c. i and iv

d. i, ii, iii, iv and v

6 Animal Cognition and Communication

This chapter contains questions about the mental lives of animals and the methods they use to communicate.

Foundation

6.1f   The study of the influence of conscious awareness and intention on an animal's natural behaviour is called

- a. comparative psychology
- b. behavioural ecology
- c. cognitive ethology
- d. neurobiology

6.2f   Complete the following sentence using one of the options below: '*Clever Hans* was a .... that appeared to be able to count.'

- a. pig
- b. horse
- c. chimpanzee
- d. jackdaw

6.3f   Complete the following definition of tool use employed by Beck (1980) using one of the options in Table 6.1:

© Paul A. Rees 2022. *Key Questions in Animal Behaviour and Welfare:*
A Study and Revision Guide (P.A. Rees)
DOI: 10.1079/9781789248975.0006

...the ..... K..... employment of an ..... L ..... environmental object to alter the form, position, or condition of another object, another organism, or the user itself when the user holds or carries the tool during or just prior to use and is ..... M ..... for the proper and effective ..... N ..... of the tool.

Table 6.1

Word	A	B	C	D
K	external	internal	external	external
L	attached	attached	unattached	unattached
M	accountable	responsible	liable	responsible
N	orientation	manufacture	location	orientation

   a. A

   b. B

   c. C

   d. D

6.4f The capacity that some animals have to impute a particular mental state to another is called

   a. mental state imputation

   b. mental state assignment

   c. mental state attribution

   d. mental state knowledge

6.5f Some apes have been taught to communicate with humans by pointing to images representing words called

   a. lexigrams

   b. emojis

   c. hieroglyphs

   d. lexicons

6.6f **Which of the following is well-known for studying self-recognition in primates?**

a. Gordon G. Gallup Jr

b. Jane Goodall

c. Robert Yerkes

d. John Napier

6.7f **Overmarking is**

a. a pattern of stripes used by some species as camouflage

b. a pattern of claw marks made on trees by some carnivorous mammals

c. the spreading of excessive quantities of dung by some mammals when marking their territories

d. scent marking over the marks of conspecifics

6.8f **Stridulation is the name of the process that produces the sounds used in communication in**

a. lemurs

b. bats

c. crickets

d. dolphins

6.9f **The process by which the receiver of a communication translates the signal into changes in its internal state is called**

a. encoding

b. decoding

c. deciphering

d. ciphering

6.10f **The sleep cycle in mammals consists of**

a. alpha sleep and delta sleep

b. active sleep and quiet sleep

c. active sleep and alpha sleep

d. beta sleep and quiet sleep

6.11f   Which of the following statements about pheromones is false?

    a. They are only present in insects

    b. They can affect behaviour in very low concentrations

    c. Some pheromones act as alarm signals to conspecifics

    d. Most sex pheromones are produced by females

6.12f   Male members of which of the following taxa wash their hands in their own urine to attract the attention of females?

    a. Tarsiers (Tarsiidae)

    b. Vervet monkeys (*Chlorocebus pygerythrus*)

    c. Capuchin monkeys (Cebidae)

    d. Tree shrews (Scandentia)

6.13f   The first pheromone to be studied was found in the

    a. Atlas moth (*Attacus atlas*)

    b. Emperor moth (*Saturnia pavonia*)

    c. Eyed hawk-moth (*Smerinthus ocellata*)

    d. Chinese silk moth (*Bombyx mori*)

6.14f   Foot-flagging is a visual signal used by some

    a. anurans

    b. lemurs

    c. songbirds

    d. rodents

6.15f   Experimental blockage of the vomeronasal system of the European adder (*Vipera berus*) prevents males from fighting with one another. This suggests that they recognise each other by

    a. acoustic cues

    b. visual cues in the visible spectrum

    c. chemical cues

    d. infrared visual clues

6.16f Wolves communicate in part by using their tails. Match the behaviours listed in Table 6.2 with the tail positions shown in Fig. 6.1

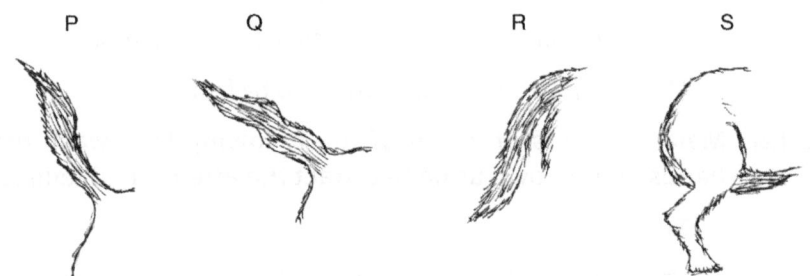

Fig. 6.1.

Table 6.2

Meaning	A	B	C	D
P	Submission	Self-confident in social situations	Normal attitude in absence of social pressure	Certain threat
Q	Self-confident in social situations	Certain threat	Self-confident in social situations	Normal attitude in absence of social pressure
R	Certain threat	Normal attitude in absence of social pressure	Certain threat	Self-confident in social situations
S	Normal attitude in absence of social pressure	Submission	Submission	Submission

    a. A

    b. B

    c. C

    d. D

6.17f **Vervet monkeys (*Chlorocebus pygerythrus*) have an alarm call that means 'snake'. If one individual in a group makes this call other group members**

a. sit down and look down

b. sit down and look up

c. stand upright and look down

d. stand upright and look up

6.18f **Bees communicate using two movement patterns to convey information about food to conspecifics: the 'round dance' and the 'waggle dance'. Which of the following statements is true?**

a. The 'round dance' conveys information about the direction of food and the distance; the 'waggle dance' tells other bees that food is near

b. The 'round dance' tells other bees that food is near; the 'waggle dance' conveys information about the direction of food and the distance

c. The 'round dance' tells other bees that food is near; the 'waggle dance' conveys information about the direction of food only

d. The 'round dance' tells other bees that food is near; the 'waggle dance' conveys information about the distance to food only

6.19f **Which of the following species uses an elaborate system of hand signals to communicate with conspecifics?**

a. Gorilla (*Gorilla gorilla*)

b. Indri (*Indri indri*)

c. Siamang (*Symphalangus syndactylus*)

d. Bonobo (*Pan paniscus*)

6.20f **Bottlenose dolphins (*Tursiops* spp.) transmit information about their individual identities by producing a sound known as**

a. a name click

b. a signature whistle

c. a name coda

d. an identity whistle

Intermediate

6.1i Which of the following scientists conducted laboratory experiments on the ability of chimpanzees to devise and use simple tools and build structures, for example, piling boxes on top of each other to create a platform so that they could reach a suspended food item?

a. Karl von Frisch

b. Burrhus F. Skinner

c. Wolfgang Köhler

d. Jane Goodall

6.2i Complete the following sentence using one of the terms listed below: '*Kanzi* was a .... who was the subject of a language study at Georgia State University in the United States and learned to communicate with humans.'

a. male bonobo (*Pan paniscus*)

b. male orangutan (*Pongo pygmaeus*)

c. female chimpanzee (*Pan troglodytes*)

d. female gorilla (*Gorilla gorilla*)

6.3i Differences in the vocalisations produced by individuals belonging to different populations of the same species are known as

a. sociolects

b. accents

c. dialects

d. idiolects

6.4i **Individuals of which of the following species communicate over long distances using their feet to create seismic waves?**

a. American bison (*Bison bison*)

b. African elephant (*Loxodonta africana*)

c. Hippopotamus (*Hippopotamus amphibius*)

d. White rhinoceros (*Ceratotherium simum*)

6.5i **The chimpanzee *Washoe* was the first non-human to learn to communicate with humans using**

a. American sign language

b. lexigrams

c. British sign language

d. a synthesised voice operated by a touch screen computer

6.6i **Some animals use immobile objects as 'tools' but do not hold them or manipulate them in any way. For example, some birds (e.g. crows (*Corvus corone*)) drop food items onto rocks to break them open. Such objects are often called**

a. primary tools

b. secondary tools

c. proto-tools

d. ancillary tools

6.7i **Reliable information transmitted from a sender (signaller) to a receiver is called**

a. a candid signal

b. a truthful signal

c. a sincere signal

d. an honest signal

6.8i **Individual vocalisations made by wolves may be distinguished using a**

a. sound spectrograph

b. sound spectrometer

c. spectroscope

d. scintillation camera

6.9i **Jays (Corvidae) store food in caches from which they retrieve food at a later date. A single bird may hide food in thousands of caches. It remembers their locations using a**

a. mental compass

b. cognitive map

c. magnetic sense

d. taxis

6.10i **When observed from above, the 'waggle dance' of a honeybee takes the form of a figure**

a. three

b. six

c. eight

d. zero

6.11i **The theory that claims that some animals are capable of mental state attribution – that is, the ability to understand what another individual is thinking – is the**

a. theory of mind

b. theory of thought

c. theory of psyche

d. theory of intellect

6.12i **In the 1980s Roger Fouts reported observations of the cultural transmission of sign language between captive**

a. bonobos (*Pan paniscus*)

b. gorillas (*Gorilla gorilla*)

c. orangutans (*Pongo pygmaeus*)

d. chimpanzees (*Pan troglodytes*)

6.13i **Which of the following animals communicate by producing click trains or codas?**

a. elephants

b. cetaceans

c. woodpeckers

d. squirrels

6.14i **Lip smacking is a form of communication used by some**

a. anurans

b. cervids

c. primates

d. canids

6.15i **Which of the following examples of tool use in birds is fictitious?**

a. The black kite (*Milvus migrans*) spreads bush fires by transporting burning sticks in its talons or beak so that it can hunt prey at the edge of the fire

b. The woodpecker finch (*Camarhynchus pallidus*) uses a cactus spine to reach insect larvae in tree crevices

c. Striated herons (*Butorides striata*) drop bread crumbs into water as fish bait.

d. None of the above is fictitious.

6.16i **African elephants (*Loxodonta africana*) communicate with each other using**

a. infrasound

b. seismic communication

c. chemical signals

d. all of the above

6.17i **The theory that animals send honest signals because dishonest signals are too costly to fake is known as the**

a. handicap principle

b. impediment principle

c. disadvantage principle

d. cost-benefit principle

6.18i **'A cooperative species-typical vocal behavior characterized by the reciprocal exchange of long-distance contact calls between conspecifics' (Miller *et al.*, 2009) is a definition of**

a. requited calling

b. antiphonal calling

c. mutual calling

d. reciprocal calling

6.19i **Some lepidopterans possess eye-spots on their wings which may frighten predators when they are suddenly revealed. This type of display is known as a**

a. deimatic display

b. disruptive display

c. distraction display

d. dimorphic display

6.20i **Which of the following statements about dreaming in mammals are true (Fig. 6.2)?**

i. It is accompanied by rapid eye movements

ii. It occurs during quiet sleep

iii. It can be identified with an electroencephalogram (EEG)

iv. In some species it is accompanied by ear movements

a. i and ii

b. i and iii

c. i, ii and iii

d. i, iii and iv

Fig. 6.2.

Advanced

6.1a   The red spot test (or mirror test) is used to test for

    a. learning

    b. self-recognition

    c. memory

    d. imprinting

6.2a   Darcin is

    a. a sex pheromone found in mice

    b. a chemical signal produced by leafcutter ants

    c. a pheromone produced by female moths

    d. a chemical component of the scent marks left by some antelopes

6.3a   Lobsters communicate with each other using pheromones released in urine and squirted from the base of their antennae. These chemicals play an important part in

    i.   maintaining a dominance hierarchy among males

    ii.   mate choice and mating behaviour

    iii. the identification of previous opponents

a. i only

b. i and iii

c. ii and iii

d. i, ii and iii

6.4a **Culturally transmitted dialects of acoustic signals are the basis of the largest affiliative structures in sperm whale (*Physeter macrocephalus*) societies. These groups are known as**

a. vocal kinship groups

b. vocal tribes

c. vocal clans

d. vocal families

6.5a **Which of the following species engages in totem tree marking: the preferential repeated scent-marking of certain trees?**

a. Pharaoh ant (*Monomorium pharaonis*)

b. Sifaka (*Propithecus* spp.)

c. Bonobo (*Pan paniscus*)

d. Naked mole rat (*Heterocephalus glaber*)

6.6a **In some species individuals occasionally produce an alarm call indicating the presence of a predator when, in fact, there is no predator present. The other members of the group run to safety, leaving the fake alarm caller to access food or other resources. The false alarm call is not ignored by the other group members because**

a. most alarm calls are genuine and ignoring an alarm call could have fatal consequences

b. the callers are always dominant individuals

c. the callers are perceived as the most alert individuals in the group

d. the call is repeated until all of the other group members respond

6.7a **An action used by a signaller to draw a receiver's attention to a specific object is called**

   a. a gesticulation

   b. an indicative gesture

   c. a referential gesture

   d. a connotative gesture

6.8a **Swartz and Evans (1991) exposed 11 chimpanzees (*Pan troglodytes*) to a mirror and tested them with the 'mark test'. From their results they concluded that**

   a. only male chimpanzees show self-recognition

   b. not all chimpanzees show self-recognition

   c. only female chimpanzees show self-recognition

   d. all chimpanzees show self-recognition

6.9a **In the firefly *Photinus pyralis*, males recognise females of the species by the duration of the flashes of light she produces. This bioluminescence is the result of the release of energy caused by the oxidation of**

   a. luciferin

   b. rhodopsin

   c. porphyropsin

   d. iodopsin

6.10a **Figure 6.3 shows a brown bear (*Ursus arctos*) attempting to remove fruit from a metal box that has a hinged door secured with a sliding latch. The only way of accessing the fruit is by sliding the latch. This experiment is designed to study**

   a. discrimination learning

   b. habituation

   c. classical conditioning

   d. trial-and-error learning

Fig. 6.3.

6.11a Which of the following taxa engage in antiphonal calling?

    i. Marmosets

    ii. Vampire bats

    iii. European starlings

    iv. Tree shrews

    v. African elephants

    a. iii and v

    b. i, iii and v

    c. i, iii iv and v

    d. i, ii, iii, iv and v

6.12a Under experimental conditions two Siamese fighting fish (*Betta splendens*) exhibited different aggressive behaviours towards each other depending upon whether they were in the presence of a male or a female conspecific prior to interacting with each other. This phenomenon is known as the

i. audience effect

ii. spectator outcome

iii. bystander effect

iv. viewer effect

a. iv

b. i or iii

c. ii or iv

d. ii

6.13a **Before playing with a cub an adult male lion (*Panthera leo*) lowers his forequarters and 'bows' to indicate that the behaviours that follow are non-aggressive. This phenomenon is known as**

a. multicommunication

b. ultracommunication

c. metacommunication

d. megacommunication

6.14a **Dufour's gland produces a chemical attractant in**

a. ants, wasps and bees

b. butterflies and moths

c. spiders

d. cockroaches

6.15a **Individuals E, F and G all belong to the same species (Fig. 6.4). When individual E attempts to communicate with F only and the signal is also received by individual G this interception of the signal by G is known as**

a. spying

b. eavesdropping

c. snooping

d. surveillance

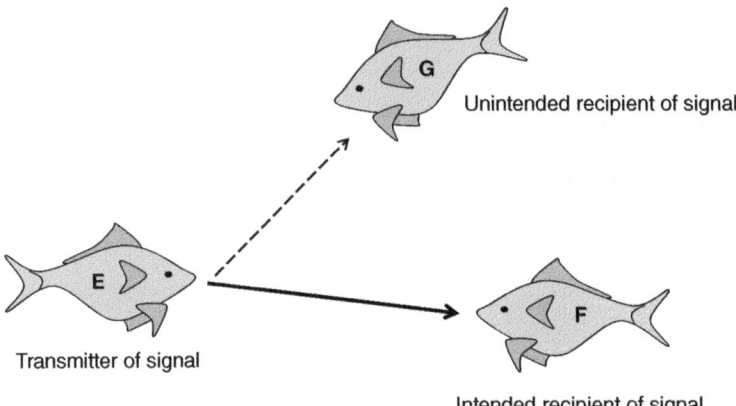

Fig. 6.4.

6.16a **Which of the following has demonstrated an ability to construct composite tools, i.e. tools made of complimentary different parts firmly connected together?**

    a. Crested jay (*Platylophus galericulatus*)

    b. Australian raven (*Corvus coronoides*)

    c. Blue jay (*Cyanocitta cristata*)

    d. New Caledonian crow (*Corvus moneduloides*)

6.17a **Which of the following statements about acoustic signals is false?**

    a. Acoustic signals propagate in air or water so they are ideally suited to communication over long distances

    b. High frequency sounds travel further than low frequency sounds

    c. In frogs, producing sounds has a high energetic cost

    d. Temperature affects the rate of acoustic signal production in frogs

6.18a **Plotnik *et al.* (2006) erected a large mirror in the enclosure belonging to *Happy*, *Maxine* and *Patty* at the Bronx Zoo in New York City as part of an experiment in mirror self-recognition. These animals were**

    a. Asian elephants (*Elephas maximus*)

    b. chimpanzees (*Pan troglodytes*)

c. orangutans (*Pongo pygmaeus*)

d. giraffes (*Giraffa camelopardalis*)

6.19a Which of the following features must a gesture possess if it is to be considered referential?

i. It must be directed towards an object or specific area of the signaller's body

ii. It must be a mechanically ineffective movement e.g. it must not be the equivalent of human pointing

iii. It must be aimed at a potential recipient

iv. It must receive a voluntary response from the recipient

v. It must possess the hallmarks of intentional production

a. i, iii and iv

b. i, ii, iii and iv

c. iii, iv and v

d. i, ii, iii, iv and v

6.20a The subterranean African demon mole rat (*Tachyoryctes daemon*) communicates using substrate-borne vibrations caused by striking the

a. floor of the tunnel with its tail

b. floor of the tunnel with its head

c. ceiling of the tunnel with its head

d. sides of the tunnel with its body

7 Behavioural Ecology and Social Behaviour

This chapter contains questions about social behaviour and the effect of evolution on producing adaptive behaviour in animals.

Foundation

7.1f   Which of the following species engage(s) in food begging?

    a. Spotted hyena (*Crocuta crocuta*)

    b. Herring gull (*Larus argentatus*)

    c. African painted dog (*Lycaon pictus*)

    d. All of the above

7.2f   Elephants and many primate species live in societies in which the size, composition and dispersion of the social groups change with time and as animals move through their environment. For example, individuals may sleep together at night but forage in small groups during the day. Such a society is a

    a. fission-fission society

    b. fusion-fusion society

    c. fission-fusion society

    d. fusion-fission society

© Paul A. Rees 2022. *Key Questions in Animal Behaviour and Welfare: A Study and Revision Guide* (P.A. Rees)
DOI: 10.1079/9781789248975.0007

7.3f   **Which of the following mammals may be described as eusocial?**

a. Tiger (*Panthera tigris*)

b. Naked mole rat (*Heterocephalus glaber*)

c. Meerkat (*Suricata suricatta*)

d. Yellow mongoose (*Cynictis penicillata*)

7.4f   **Which of the following suggested that social behaviours, especially social signals, in animals are important in population regulation?**

a. William Hamilton

b. Edward O. Wilson

c. Herbert Andrewartha

d. Vero Wynne Edwards

7.5f   **Complete the following sentence using one of the terms listed below: 'Lowering the head, rolling over and exposing the underside of the body or throat, food begging and presenting the buttocks are all .... behaviours in mammals.'**

a. aggressive

b. submissive

c. assertive

d. courtship

7.6f   **A study by Froy *et al.* (2016) found that in red deer (*Cervus elephus*) rearing a male calf results in**

a. an increased risk of mortality in the mother

b. no change in the risk of mortality in the mother

c. a decreased risk of mortality in the father

d. an increase in the probability that the mother's next calf will be female

7.7f **The type of natural selection that considers the role of relatives when assessing the genetic fitness of an individual is called**

a. brood selection

b. clan

c. kin selection

d. family selection

7.8f **Which of the following statements about giraffes (*Giraffa camelopardalis*) is true?**

a. They rear their offspring in nursery groups

b. They live in fission-fusion social systems

c. Females with offspring maintain stronger associations than those without offspring

d. All of the above are true

7.9f **The seasonal sexual activity of mammals such as deer, goats, camels, bison and moose is known as**

a. lekking

b. rutting

c. musth

d. tupping

7.10f **The behaviour whereby, after copulation, one individual of a pair prevents the other from seeking additional mates or prevents additional mates from approaching is called**

a. pair bonding

b. mate defending

c. mate shielding

d. mate guarding

7.11f **The time and resources an individual animal must spend to produce and raise offspring is known as**

a. maternal cost

b. motherly expenditure

c. parental investment

d. parenting outlay

7.12f The mating system in which a single male mates with many females is known as

a. monogamy

b. polygyny

c. polygamy

d. polyandry

7.13f Which of the following would you expect to be correlated with genetic fitness in gemsbok (*Oryx g. gazella*)?

a. Horn symmetry

b. Hoof size

c. Tail length

d. Coat colour

7.14f In social insects drones, soldiers and workers belong to different

a. tribes

b. clones

c. castes

d. ranks

7.15f An area where the males of some bird species (e.g. grouse) gather to engage in competitive displays and courtship rituals is called a

a. lekvar

b. lekach

c. lekane

d. lek

7.16f A female who assists a mother in the rearing of her young is most accurately known as

a. a pseudo-mother

b. an allomother

c. an aide

d. a worker

7.17f **An animal that exploits the parental care of individuals that are not its parents, e.g. a cuckoo chick (*Cuculus canorus*) receiving care from a pair of reed warblers (*Acrocephalus scirpaceus*), is called a**

a. nest parasite

b. brood parasite

c. clutch parasite

d. roost parasite

7.18f **Where do male lyrebirds perform their courtship displays?**

a. On circular mounds of bare soil on the forest floor

b. In the air

c. On platforms made of twigs built in large trees

d. On lakes and other bodies of freshwater

7.19f **A male dunnock or hedge sparrow (*Prunella modularis*) will peck at the cloaca of a female**

a. to persuade her to mate

b. to cause her to eject the sperm of a rival male

c. as a sign of aggression

d. immediately after mating with her

7.20f **Which of the following illustrates the relationships between individuals in a linear hierarchy? (Note: A>B means A is dominant to B).**

a. G>B>F>D

b. K<H>L

c. C<B>D<A

d. J>D>F<G

Intermediate

7.1i **The theory that individuals in a population try to reduce their predation risk by putting conspecifics between themselves and predators is known as the**

    a. anti-predator theory

    b. aggregation theory

    c. selfish herd theory

    d. altruistic herd theory

7.2i **The hypothesis that it makes more evolutionary sense for an older female mammal to concentrate on rearing existing young and ensuring their survival rather than risk age-related death while giving birth in an attempt to leave more offspring is called the**

    a. mother hypothesis

    b. parental investment hypothesis

    c. unproductive senescence hypothesis

    d. optimal fitness hypothesis

7.3i **In African painted dogs (*Lycaon pictus*) (Fig. 7.1)**

    a. there is no dominance hierarchy

    b. only males exhibit a dominance hierarchy

    c. only females exhibit a dominance hierarchy

    d. males and females have separate dominance hierarchies

Fig. 7.1.

7.4i **Many species have evolved cooperative feeding behaviours. Which of the following species uses bubble nets to catch food in the ocean?**

a. Humpback whale (*Megaptera novaeangliae*)

b. Blue whale (*Balaenoptera musculus*)

c. Killer whale (*Orcinus orca*)

d. Narwhal (*Monodon monoceros*)

7.5i. **Appeasement behaviour in terrestrial mammals often involves**

a. raising the head and body

b. lowering the head and body

c. a head shaking movement

d. raising the tail

7.6i **Individuals of some fish species are able to change sex under certain conditions. In many species one sex is larger than the other. Which of the following scenarios would make the ability to change sex advantageous?**

a. Recovery from overfishing where larger individuals are targeted by fishermen

b. Male-to-female sex change where females benefit more than males from being larger because they can produce more eggs

c. Female-to-male sex change where males gain more from being large because they are better able to defend territories and mate with many females

d. All of the above

7.7i **The abbreviation ESS stands for**

a. environmentally stable strategy

b. evolutionarily strategic state

c. ecologically static strategy

d. evolutionarily stable strategy

7.8i **How does a female infant olive baboon (*Papio anubis*) (Fig. 7.2) derive her social rank?**

a. From that of her mother by inheritance

b. From that of her father by inheritance

c. As a result of contests with others

d. She has no rank

Fig. 7.2.

7.9i **A 'faeder' is**

a. a male ruff (*Philomachus pugnax*) that mimics the appearance of a female to gain access to females

b. a non-breeding hartebeest (*Alcelaphus buselaphus*)

c. a dominant male Chacma baboon (*Papio ursinus*)

d. a social spider that cooperates with others to catch prey in a communal web

7.10i **During the rutting season adult red deer (*Cervus elephus*) stags (Fig. 7.3) engage in roaring contests whose purpose is to**

a. allow rival stags to locate each other

b. allow rival stags to assess each other's fighting ability

c. attract hinds

d. deter subadult stags from stealing hinds

Fig. 7.3.

7.11i **'Helpers at the nest' – juveniles and sexually mature adolescents (of either sex) that remain with their parents to assist in the rearing of future offspring – occur in**

    i. birds

    ii. meerkats (*Suricata suricatta*)

    iii. Damaraland mole rats (*Fukomys damarensis*)

    iv. carpenter bees (Xylocopinae)

    a. i only

    b. i and ii

    c. i, iii and iv

    d. i, ii, iii and iv

7.12i **The coefficient of relatedness (r) for different classes of kin specify the probability of sharing a particular allele by** common descent (Table 7.1). What would be the probability of an allele for altruism being passed from an uncle to his niece and then from her to her son.

Table 7.1

Relationship	Coefficient of relatedness (r)
Parent-offspring	0.50
Siblings	0.50
Non-relatives	0
Grandparent-grandchild	0.25
Uncle/aunt-nephew/niece	0.25

a. 0.75

b. 0.25

c. 0.125

d. 0.5

7.13i Whether or not a parent is willing to invest time and resources in the rearing of offspring is, at least in part, determined by whether or not they can be certain that particular animals are definitely *their* offspring. Those female animals that give birth to their young are generally able to achieve

a. parental certainty

b. maternal assurance

c. parental confidence

d. maternal conviction

7.14i The male of some spider species presents his mate with a 'nuptial gift' to prevent her from

a. resisting copulation

b. eating her own eggs

c. eating him

d. mating with a different male

7.15i **In which of the following animals are older individuals important repositories of social knowledge?**

a. African elephants (*Loxodonta africana*)

b. Giraffes (*Giraffa camelopardalis*)

c. Cetaceans (Cetacea)

d. All of the above

7.16i **The mating system whereby males that mate with more than one female monopolise resources needed by receptive females is called**

a. resource defence polyandry

b. lek polyandry

c. resource defence polygyny

d. lek polygyny

7.17i **Which of the following expressions accurately reflects the relationship between the different types of fitness?**

a. Direct fitness = Inclusive fitness + Indirect fitness

b. Indirect fitness = Inclusive fitness + Direct fitness

c. Inclusive fitness = Direct fitness + Indirect fitness

d. Indirect fitness = Inclusive fitness × Direct fitness

7.18i **When two competing individuals of the same species do not have the same choice of strategies or prospective payoffs in a contest (e.g. a territory owner versus an intruder) it is said to be**

a. an unfair contest

b. an asymmetric contest

c. a biased contest

d. a prejudiced contest

7.19i **Brood parasitism occurs in which of the following taxa?**

a. Birds

b. Fishes

c. Insects

d. All of the above

7.20i The Fraser Darling effect is most likely to be observed in animals that breed

a. every other year

b. on cliffs

c. in large groups

d. in very small groups

Advanced

7.1a According to John Maynard Smith (Smith, 1984), what is defined as:

...a method of analysing the evolution of phenotypes (including types of behaviour) when the fitness of a particular phenotype depends on its frequency in the population?

a. Ecological set theory

b. The prisoner's dilemma

c. Evolutionary game theory

d. Selfish gene theory

7.2a The mating system whereby males that mate with more than one female are non-territorial and seek out scattered receptive females is called

a. female defence polygyny

b. scramble competition polygyny

c. scramble competition polyandry

d. female defence polyandry

7.3a The hypothesis that states that females receive sperm from several males to allow sperm competition for their eggs and thus 'weed out' genetically inferior sperm is known as the

a. competing sperm hypothesis

b. superior sperm hypothesis

c. rival sperm hypothesis

d. good sperm hypothesis

7.4a In elephants, a 'mating pandemonium' involves

a. a male and female only, before courtship begins

b. a male and female only at the end of courtship

c. a male and female before copulation plus other members of the herd

d. a male and female after copulation plus other members of the herd

7.5a Animal X sometimes helps animal Y. Animal Y sometimes helps animal X. In doing this each incurs a cost. This relationship is most accurately called

a. reciprocal altruism

b. altruism

c. reciprocal cooperation

d. social altruism

7.6a A predator feeds by chasing and catching its prey. It may choose a prey organism that is large or one that is small and it may chase it at high speed or at low speed. The feeding strategy of a predator, in terms of the costs and benefits of chasing large and small prey at high and low speeds, is best assessed by calculating

a. a game matrix

b. a correlation coefficient

c. an activity budget

d. a predation index

7.7a Which of the following statements about spotted hyenas (*Crocuta crocuta*) (Fig. 7.4) is false?

a. Females possess an enlarged clitoris known as a pseudopenis

b. The sons of alpha females 'inherit' their mother's high dominance status

c. Males and females form long-term pair bonds

d. Hyenas engage in elaborate greeting ceremonies

Fig. 7.4.

7.8a **The analysis of the potential costs and benefits of decisions about whether individual animals should cooperate or not towards a particular end is considered in the game known as the**

    a. hostage quandary

    b. prisoner's dilemma

    c. traitor's problem

    d. altruist's dilemma

7.9a **Natural selection will favour the evolution of altruism when**

$rb - c > 0$

where,

c = the fitness cost to the altruist

b = the fitness benefit to the beneficiary

r = the coefficient of relatedness between the altruist and the beneficiary

This expression is known as

    a. Wilson's rule

    b. Hamilton's rule

    c. Haldane's rule

    d. Simpson's rule

7.10a The circles in Fig. 7.5 represent the positions of individual prey animals in a herd seen from above. According to Hamilton's selfish herd theory, which position in the group provides the best protection from predators?

   a. A

   b. B

   c. C

   d. D

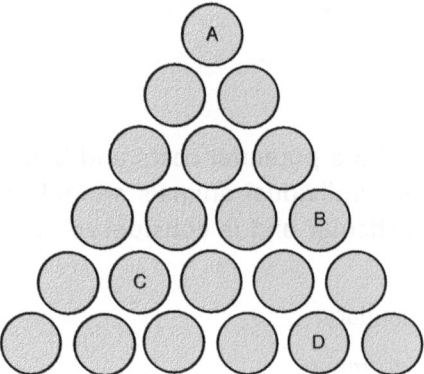

Fig. 7.5.

7.11a The term 'pecking order' was first used when discussing dominance in

   a. turkeys

   b. parakeets

   c. crows

   d. domestic fowl

7.12a The hypothesis that host birds only tolerate a brood parasite's eggs in their nests when they fear retaliation by the parasite if these eggs are removed in the form of depredation of their own eggs is called the

   a. retaliation hypothesis

   b. mafia hypothesis

   c. protection racket hypothesis

   d. revenge hypothesis

7.13a The mating of a low-ranking male with a female in a group to whom he would not normally have access – because of his low rank – is known as a

   a. cunning mating

   b. lucky mating

   c. sneaky mating

   d. devious mating

7.14a Black-headed gulls (*Chroicocephalus ridibundus*) are ground-nesting seabirds that pick up and carry away egg shells shortly after their chicks have hatched (Tinbergen, 1963). Kittiwakes (*Rissa tridactyla*) nest on narrow cliff edges and do not remove egg shells (Cullen, 1957). These observations support the hypothesis that egg shell removal is

   a. a behaviour that contributes to nest hygiene

   b. unlikely to affect egg survival

   c. an anti-predator behaviour

   d. not adaptive

7.15a The percentage of various classes of individual barn swallows (*Hirundo rustica*) (Fig. 7.6) present in colonies and participating in the mobbing of potential predators are shown in Table 7.2. Mobbing occurs primarily during the breeding season. Several hypotheses have been suggested to explain mobbing in this species (Shields, 1984) (Table 7.3)

Fig. 7.6.

Table 7.2

Class of swallow	Percentage of population	Percentage of active mobbers
Adult unmated	6	2
Adult before incubation	9	11
Adult during incubation	14	10
Adult with young	51	77
Juvenile	20	1

Table 7.3

Hypothesis	Prediction
i. Self-defence	There should be no seasonal variation in mobbing
ii. Mating advertisement	Mobbing will occur in the breeding season and unmated adults will be well-represented in the population of mobbers
iii. Parental care	Mobbers should primarily be breeding adults with young to defend

The hypothesis most likely to be true based on the evidence provided here is

a. i

b. ii

c. iii

d. a combination of i and ii

7.16a The insurance egg hypothesis is a possible explanation for the evolution of

a. twins in mammals

b. obligate siblicide in birds

c. parthenogenesis in lizards

d. mouth brooding in fishes

7.17a In an asymmetric contest where there are differences in fighting ability in which individuals A and B can vary their cost of escalation along a continuum, an evolutionarily stable strategy (ESS) will evolve such that A will give up and retreat when

i. $\dfrac{V_A}{K_A} > \dfrac{V_B}{K_B}$

ii. $\dfrac{V_A}{K_A} = \dfrac{V_B}{K_B}$

iii. $\dfrac{V_A}{K_A} < \dfrac{V_B}{K_B}$

iv. $\dfrac{V_A}{K_A} = 1$

where,

V_A = the value of the resource to A

V_B = the value of the resource to B

K_A = the rate at which costs are accrued during the contest by A

K_B = the rate at which costs are accrued during the contest by B

a. i

b. ii

c. iii

d. iv

7.18a In Q7.17a V, the value of the resource to the contestants, is best described as the

a. increase in the number of offspring produced by the contestant as a result of obtaining the resource

b. increase in Darwinian fitness of an individual as a result of obtaining the resource

c. increase in the number of the contestant's offspring that survive as a result of obtaining the resource

d. increase in the social status of the contestant as a result of obtaining the resource

7.19a **Which of the following is not required for reciprocal altruism to evolve in an animal society?**

   a. The animals must have a short lifespan

   b. The individuals must be able to recognise each other

   c. The cost of the altruistic act must be lower than the benefit to the actor so there is a net gain over time

   d. A mechanism must exist for the detection of 'cheaters'

7.20a **The theory that the behavioural characteristics of an animal offer the best ratio of fitness benefits to fitness costs is called**

   a. operational theory

   b. selection theory

   c. optimality theory

   d. direct fitness theory

8 Measuring, Recording and Analysing Animal Behaviour and Welfare

This chapter contains questions about the methods and equipment used to record and analyse behaviour, the assessment of animal welfare and the use of animals in laboratory experiments.

Foundation

8.1f A condition scoring system may be used to monitor

 a. the numerical performance score achieved in a test by an animal that has been the subject of operant conditioning

 b. the performance of apes learning to communicate with people by pointing at pictures

 c. the level of cortisol in the blood of a large mammal

 d. the physical (body) condition of an animal for the purpose of assessing its welfare

8.2f The chimpanzees (*Pan troglodytes*) of the Gombe National Park in Tanzania have been studied for over 60 years. Such a study, examining the same group of animals over an extended a period of time, is called a

 a. cross-sectional study

 b. controlled study

 c. longitudinal study

 d. meta-analysis

© Paul A. Rees 2022. *Key Questions in Animal Behaviour and Welfare: A Study and Revision Guide* (P.A. Rees)
DOI: 10.1079/9781789248975.0008

8.3f In a study of free-ranging orangutans (*Pongo pygmaeus*) a researcher selected an individual at random and followed her for an hour, recording which foods she selected from the forest. At the end of the hour the researcher selected a different individual and made similar recordings. This method of sampling is called

a. scan sampling

b. focal sampling

c. one-zero sampling

d. *ad libitum* sampling

8.4f Which of the following techniques may be used to slow down the behaviour of an animal so that it may be studied in detail?

a. Stop motion photography

b. High-speed photography

c. Slow-speed photography

d. Phase contrast photography

8.5f It is increasingly common for scientists to use untrained observers to collect data on the behaviour and welfare of animals, such as zookeepers and other animal caregivers and, sometimes, members of the general public. In a study of the response of gorillas (*Gorilla gorilla*) living in zoos to veterinary treatment their keepers were asked to complete a form containing several questions for each gorilla in their care including the following:

'How would you describe the behaviour of Gorilla X when he/she is required to provide a sample of blood?

i. Always uncooperative

ii. Sometimes uncooperative

iii. Usually cooperative

iv. Always cooperative'

This question is an a example of

a. a closed question in a verbally-administered questionnaire

b. an open question in a self-administered questionnaire

c. a closed question in a self-administered questionnaire

d. a open question in a verbally-administered questionnaire

8.6f   Very rare behaviours are best recorded by

a. *ad libitum* sampling

b. scan sampling

c. focal sampling

d. extemporary sampling

8.7f   In a study of a group of meerkats (*Suricata suricatta*) (Fig. 8.1) which of the following methods of identifying individual animals is not acceptable?

a. Giving them names normally associated with humans

b. Assigning them numbers

c. Assigning them letter codes

d. All of these methods are acceptable

Fig. 8.1.

8.8f **An experiment in which animals are allowed free access to two or more different environments (e.g. rubber and concrete floors in their housing) and the amount of time spent in each is measured is known as a**

a. selection test

b. preference test

c. predilection test

d. inclination test

8.9f **The Open Field Maze was initially developed in 1934 to measure**

a. emotionality in rodents

b. cognition in crows

c. memory in marmosets

d. social behaviour in cats

8.10f **The 'five domains' model for assessing animal welfare was proposed by**

a. Mellor and Reid (1994)

b. Kirkwood and Mason (1994)

c. Broom and Clubb (1994)

d. Dawkins and Bekoff (1994)

8.11f **When an experimental animal is subjected to 'pithing'**

a. it is deprived of oxygen

b. its brain is removed

c. it spinal cord is removed

d. its spinal cord is severed

8.12f **For the purpose of recording, behaviours may be divided into 'states' that have an appreciable duration and 'events' that are instantaneous. Match the behaviours listed in Table 8.1 with the correct category.**

Table 8.1

Behaviour	A	B	C	D
Biting a conspecific	Event	Event	Event	Event
Sleeping	State	State	Event	State
Feeding	State	State	State	Event
A copulation	Event	Event	State	Event
Suckling	State	Event	State	State

    a. A

    b. B

    c. C

    d. D

8.13f **In a study of the welfare of farmed pigs a researcher weighed 500 individual female piglets at the age of 6 months. Each of these weights was a**

    a. repeat

    b. duplicate

    c. replicate

    d. pseudoreplicate

8.14f **A Skinner box is used in the study of**

    a. learning

    b. imprinting

    c. fixed action patterns

    d. instinct

8.15f **An actograph is**

    i. a device that may be used to monitor and record any of a number of types of movements of animals

    ii. an obsolete device for recording muscle movements

    iii. a type of maze used in learning experiments with small mammals

    iv. a graph showing the activity budget of an animal

a. i

b. i and ii

c. iii

d. iv

8.16f **Which of the following is a directional device used in conjunction with a microphone to record animal vocalisations in the field?**

a. A parabolic detector

b. A conical reflector

c. A conical detector

d. A parabolic reflector

8.17f **The diagram below shows the distribution of woodlice in a piece of apparatus when they were given the choice between an area of high humidity and an area of low humidity (Fig. 8.2). The apparatus is known as**

a. a Petri disk

b. a selection chamber

c. a choice chamber

d. a Petri dish

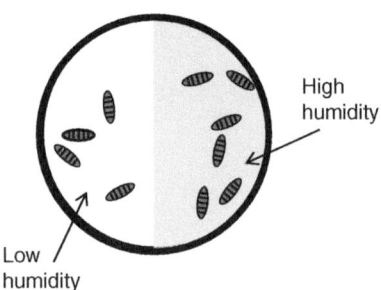

High
humidity

Low
humidity

Fig. 8.2.

8.18f **Spatial proximity loggers are used to**

a. record stress in individual animals

b. calculate the distance travelled during migration

c. study animal social networks

d. determine the genetic relationships between animals

8.19f Which of the following statements about a sociogram is false?

a. It consists of lines connecting animals that have been seen together

b. It cannot be used for large groups

c. It is based on calculations of an association index for each pair of animals

d. It is used to illustrate relationships within a group of animals

8.20f The magnitude and speed of transmission of nerve impulses can be measured using a pair of microelectrodes and a

a. mass spectrometer

b. cathode ray oscilloscope

c. spectroscope

d. digital ammeter

Intermediate

8.1i A scientist studied the activity budget of a black rhinoceros (*Diceros bicornis*). She recorded the behaviour of a single individual every 5 minutes for 3 hours. During this time the rhinoceros was recorded feeding on 23 occasions. On 8 occasions no recording could be made because the animal was out of sight. The percentage of time spent feeding was approximately

a. 23%

b. 64%

c. 42%

d. 82%

8.2i During a study of a group of ten chimpanzees (*Pan troglodytes*) a scientist recorded the type of behaviour in which each animal was engaging at 5 minute intervals from a list of

behaviours that had previously been compiled into an ethogram. This type of sampling is best described as

a. ad hoc sampling

b. random sampling

c. instantaneous scan sampling

d. fixed interval sampling

8.3i   A simulation of animal behaviours that uses repeated random sampling to generate numerical results and calculate probabilities of events that cannot be generated any other way is known as a

a. Monte Carlo simulation

b. Montessori simulation

c. Monte Forte simulation

d. Monte Bello simulation

8.4i   The apparatus in Fig. 8.3 consists of a circular channel wide enough to accommodate a small animal. During an experiment the object labelled 'X' can be moved inside the central section (which has a transparent wall) while the subject is free to move around the circular channel. This apparatus has been used to conduct experiments on

a. classical conditioning

b. imprinting

c. operant conditioning

d. insight learning

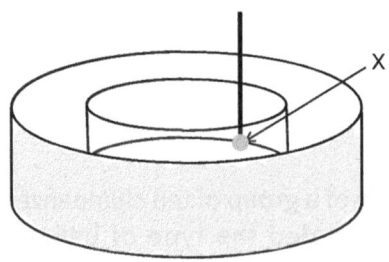

Fig. 8.3.

8.5i **The interval of time elapsing between a stimulus and response is known as the response**

a. interval

b. lag

c. latency

d. interlude

8.6i **A description of a behavioural system or process that uses mathematical language and concepts and allows predictions to be made about future events is called**

a. a template

b. a prototype

c. a facsimile

d. a model

8.7i **The apparatus shown in Fig. 8.4 was used to study**

a. feeding behaviour in marmosets by Thorpe

b. parental attachment in monkeys by Harlow

c. cognition in chimpanzees by de Waal

d. facial recognition in jackdaws by Lorenz

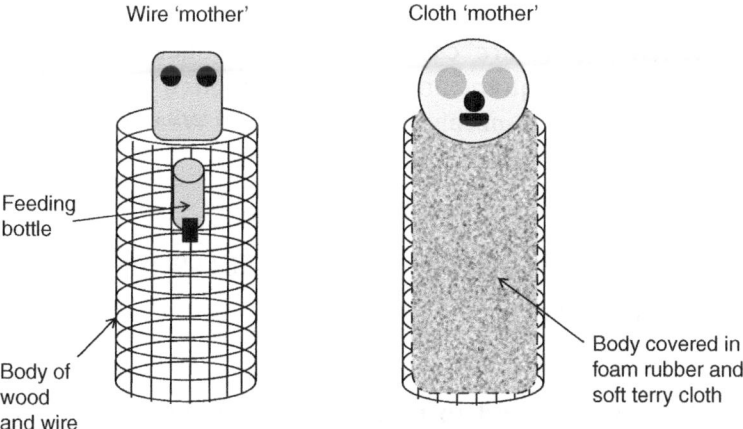

Fig. 8.4.

8.8i The laboratory test used to determine the dose or concentration of a pesticide required to kill half of the individuals exposed to it is called the

a. DL_{50} test

b. KD_{50} test

c. LD_{50} test

d. LC_{50} test

8.9i In some species, measurement of eye temperature is used as an indicator of

a. brain activity

b. oestrus state

c. mental capacity

d. welfare

8.10i Which of the following statements is true?

a. Animal welfare can be 'measured' but not 'assessed'

b. Animal welfare can be 'assessed' but not 'measured'

c. The compassionate conservation movement acknowledges that animals sometimes need to be killed for conservation purposes

d. 'Compassionate conservation' and 'conservation welfare' are different terms for the same thing

8.11i When piglets are castrated they produce 'screams' that are more frequent and of different pitch compared with when they are merely held (Taylor and Weary, 2000). These vocalisation can be considered to be

a. unreliable indications of poor welfare

b. 'honest' inter-species signals indicating poor welfare

c. dishonest signals

d. undependable signals

8.12i Studies designed to measure the relative strength of an animal's motivation to obtain different food items or other resources are called

a. consumer demand tests

b. consumer satisfaction tests

c. resource acquisition tests

d. resource allocation tests

8.13i A sociability chamber is used to study socialisation in small animals (e.g. mice and rats) and consists of

a. two chambers

b. three chambers

c. four chambers

d. five chambers

8.14i The time a mouse takes to locate and enter the escape hole is a measurement used in experiments involving a

a. T-maze

b. Morris water maze

c. Y-maze

d. Barnes maze

8.15i Which of the following lists (A–D) accurately identifies the elements of the 'five domains' model used in animal welfare assessment (Table 8.2)?

Table 8.2

A	B	C	D
Nutrition	Nutrition	Personality	Wellbeing
Temperature	Environment	Nutrition	Health
Health	Health	Wellbeing	Behaviour
Personality	Behaviour	Behaviour	Environment
Tameness	Mental state	Environment	Nutrition

a. A

b. B

c. C

d. D

8.16i **The expression 'Clever Hans effect' denotes the danger of experimenters**

    a. giving unintentional cues indicating the desired behaviour by the questioner in poorly designed cognition experiments

    b. giving anthropomorphic explanations of behaviour if their animal subjects have been given human names

    c. assuming correlation implies causation in examining the relationship between environmental variables and animal behaviour

    d. using domesticated animals in scientific experiments

8.17i **By measuring aggressiveness, exploration, activity, sociability and boldness it is possible to quantify an animal's**

    a. intelligence

    b. personality

    c. mood

    d. sentience

8.18i **A group of 10 Sulawesi crested macaques (*Macaca nigra*) (Fig. 8.5) was observed for a series of sampling periods of 15-minutes. For each period feeding behaviour was recorded as either occurring or not occurring in each individual, regardless of the number of times each animal performed this behaviour. This sampling method is known as**

    a. *ad libitum*

    b. one-zero

    c. sequence sampling

    d. scan sampling

Fig. 8.5.

8.19i A researcher produced a graph showing the changes in the amount of time a serval (*Leptailurus serval*) spent performing stereotypic pacing behaviour in a zoo enclosure using the axes in Fig. 8.6. Which of the following terms describe the variable on the y-axis (time spent pacing)?

a. Continuous and independent

b. Discontinuous and dependent

c. Discontinuous and independent

d. Continuous and dependent

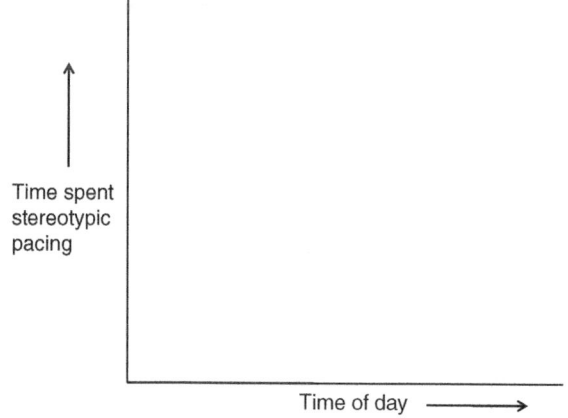

Fig. 8.6.

8.20i **Four observers collected data for a study of the activity budgets of 25 sheep (*Ovis aries*) on a farm. Only one person made the recordings on any particular day. At the outset of the study all four observers independently collected the same data at the same time for a single day and the results they obtained were compared. The purpose of this was to ensure**

a. intra-observer reliability

b. contra-observer reliability

c. infra-observer reliability

d. inter-observer reliability

Advanced

8.1a **Sometimes data collected at the beginning of a behaviour study may have to be excluded from the final analysis because the recorder has allowed the definitions of some behaviours to vary with time. This phenomenon is called**

a. observer inconsistency

b. observer error

c. observer drift

d. observer variation

8.2a **Which of the following statements about ethograms is false?**

a. An ethogram must explain the purpose of each behaviour described

b. There is no standard ethogram available that should be used for all studies of chimpanzees

c. An ethogram is a description of the behaviours exhibited by a particular species

d. An ethogram may contain a description of a limited range of behaviours, e.g. only those behaviours associated with aggression

8.3a **In a group of 4 animals how many different associations in pairs (dyads) are possible?**

a. 6

b. 12

c. 8

d. 2

8.4a A simple association index between individuals in pairs may be calculated as:

$$\frac{2N}{n_1 + n_2}$$

where,

N = the number of times animals 1 and 2 were seen together (including in a group with others)

n_1 = the total number of times animal 1 was seen (alone or as part of a group)

n_2 = the total number of times animal 2 was seen (alone or as part of a group).

If the index of association between two animals is 0.23 this means that they have been seen together

a. 23 times

b. on 23% of all study days

c. fewer times than they have been seen apart

d. as often as they have been seen apart

8.5a Which of the following statements about the association index used to study associations between animals described in Q8.4a is false?

a. The index may only be used to calculate associations between pairs of individuals

b. If it is 0 the animals have never been seen together

c. If it is 0.5 the animals have been seen together for 50% of the time

d. The highest value that the index can have is 100.0

8.6a A payoff matrix is a device used when analysing the decisions made by animals during contests using

a. group theory

b. number theory

c. set theory

d. game theory

8.7a   J. Maynard Smith defined a 'strategy' as a

a. behavioural phenotype

b. behavioural genotype

c. behavioural genome

d. meme

8.8a   The apparatus illustrated in Fig. 8.7 is called a

a. Wallis maze

b. Barnes maze

c. Morris maze

d. Butler maze

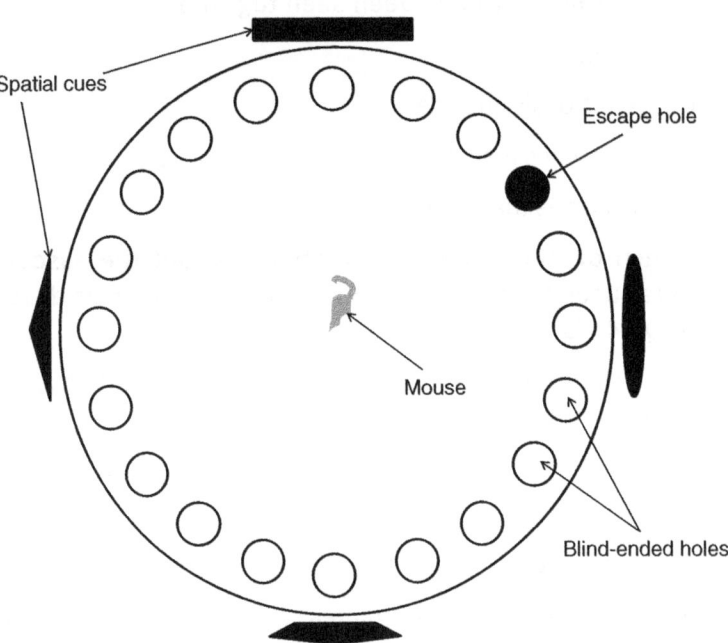

Fig. 8.7.

8.9a **Which of the following is a cost-benefit analysis for animal research in relation to the medical benefits of the work?**

a. Skinner's sphere

b. Watson's triangle

c. Bateson's cube

d. Thorndike's rectangle

8.10a **A variable that distorts the relationship between the independent and dependent variables in an animal behaviour experiment is called a**

a. compounding variable

b. confounding variable

c. compounded variable

d. extraneous variable

8.11a **Which of the following are important features of an Open Field Maze?**

i. It must be enclosed by a wall that is high enough to prevent the subject's escape

ii. It is typically square or round

iii. The area should be large enough to elicit a feeling of openness in the centre of the maze

a. i and ii

b. ii and iii

c. i and iii

d. i, ii and iii

8.12a **The Dawkins Organ is**

a. a device for recording behaviours using a keyboard

b. a device for making digital recordings of animal vocalisations

c. an electronic tool for tracking animal movements

d. a mechanical device for recording animal movements in the laboratory

8.13a **The 'five domains' model of animal welfare was an improvement over the 'five freedoms' model because the former**

   a. focusses on overall mental state

   b. emphasises that for every physical welfare compromise experienced by an animal there may be an accompanying subjective experience or emotion

   c. emphasises the importance of animals having positive experiences rather than merely the absence of negative experiences

   d. does all of the above

8.14a **A guide to the sampling methods used in the direct observation of spontaneous social behaviour in groups of animals was published in the journal *Behaviour* in 1974 by**

   a. Robert Hinde

   b. Konrad Lorenz

   c. Jeanne Altmann

   d. David Lack

8.15a **Which of the following is most likely to be exposed to the experimental procedure known as 'clock-shifting'?**

   a. A dog during a study of communication

   b. An elephant during a study of self-recognition

   c. A homing pigeon during a study of migration

   d. A monkey during a study of aggression

8.16a **The apparatus in Fig. 8.8 is a box divided into two compartments separated by a low barrier. The floor of compartment 1 can be electrified and can deliver a mild electric shock to any animal standing on it. The floor of compartment 2 cannot be electrified. At the beginning of the procedure an animal (e.g. a rat) is place on the floor in compartment 1. To avoid being shocked the animal must move to compartment 2. Prior to the shock a light illuminates for a few seconds. This apparatus is used to study avoidance learning and is called a**

   a. avoidance box

   b. shuttle box

c. transfer box

d. shunt box

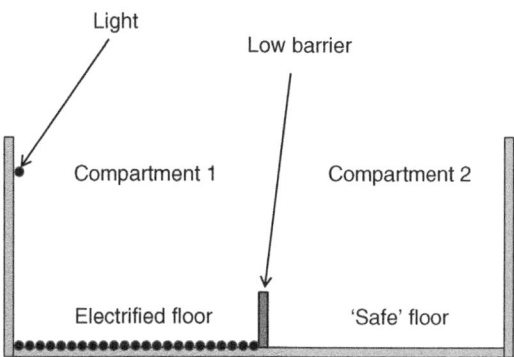

Fig. 8.8.

8.17a In the United Kingdom under the Animals (Scientific Procedures) Act 1986 a licence is required for animal experiments involving any 'protected animal'. A 'protected animal' is defined under s.1(1) of the Act as

a. any living mammal or bird

b. any living vertebrate

c. any living vertebrate other than man

d. any living vertebrate, other than man, and any living cephalopod

8.18a In the United Kingdom under the Animals (Scientific Procedures) Act 1986 a 'regulated procedure' (i.e. one that requires a Home Office licence) means any procedure 'which may have the effect of causing the animal a level of pain, suffering, distress or lasting harm equivalent to, or higher than, that caused by

a. the administration of food through a feeding tube in accordance with good veterinary practice'

b. the introduction of a needle in accordance with good veterinary practice'

c. the administration of an anaesthetic to facilitate surgery in accordance with good veterinary practice'

d. capturing the animal in a net or live trap'

8.19a The value of a commodity to an animal can be assessed by measuring the amount of work it will do to obtain it. Fig. 8.9 is a demand curve for three commodities offered to pigs. In the experiment on which this graph is based, pigs performed work – consisting of pressing a bar with the snout – to access each of the commodities. Which of the following statements is false?

a. Pigs valued food more than social contact

b. Pigs valued social contact more than sand

c. Pigs valued sand less than food

d. Pigs valued social contact less than sand or food

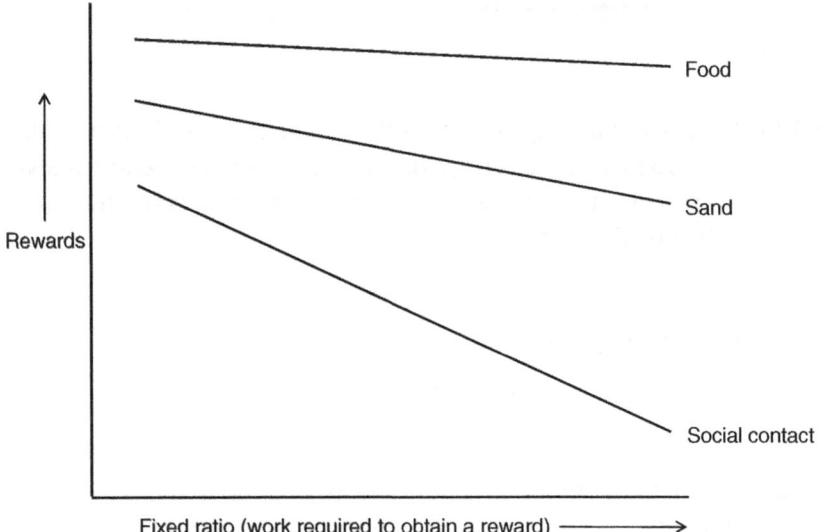

Fig. 8.9.

8.20a In Fig. 8.9 the relationship between the number of rewards obtained and the work required to obtain access to social contact showed

a. a negative correlation

b. a positive correlation

c. an exponential correlation

d. no correlation

148

9 Animal Exploitation and Welfare

This chapter contains questions about the welfare compromises that result from the various ways in which humans exploit animals, animal housing, environmental enrichment, animal diseases and the sociology of animal cruelty.

Foundation

9.1f   The term 'welfare' can relate to

    a. individual animals only

    b. animal populations

    c. animal species

    d. vertebrate species only

9.2f   Which of the columns in Table 9.1 lists the 'five freedoms'?

Table 9.1

	A	B	C	D
Freedom	From boredom	From hunger and thirst	From pain, injury and disease	From pain, injury and disease
	From hunger and thirst	From discomfort	To express normal behaviour	From boredom
	From discomfort	From pain, injury and disease	From fear and distress	From hunger and thirst
	From pain, injury and disease	To express normal behaviour	From boredom	From discomfort
	From fear and distress	From fear and distress	From hunger and thirst	To express normal behaviour

© Paul A. Rees 2022. *Key Questions in Animal Behaviour and Welfare: A Study and Revision Guide* (P.A. Rees)
DOI: 10.1079/9781789248975.0009

a. A

b. B

c. C

d. D

9.3f   Who was 'Humanity Dick'?

a. A veterinary surgeon

b. A scientist who studied animal welfare

c. A Member of Parliament in the United Kingdom

d. A clergyman who fought for animal rights

9.4f   Which of the following argued that animals are machines without feeling?

a. René Descartes

b. Lewis Gompertz

c. Frances Power Cobbe

d. Jeremy Bentham

9.5f   Humphrey Primatt published his book _A Dissertation on the Duty of Mercy and Sin of Cruelty to Brute Animals_ in 1776. Primatt was

a. a Scottish scientist

b. a Welsh surgeon

c. an English clergyman

d. an Irish politician

9.6f   The removal of part of a dog's tail for cosmetic reasons is known as

a. lopping

b. docking

c. clipping

d. paring

9.7f   Which of the following are necessary to achieve good animal welfare?

i.   A reduction in the number of animals used in research

ii. The humane treatment of animals

iii. The provision of suitable housing for animals

iv. The humane euthanasia and slaughter of animals

v.   The recognition that animals have moral rights

vi. The provision of correct nutrition for animals

a. ii, iii, iv, v and vi

b. i, ii, iii, iv and vi

c. i, ii, iii and iv

d. i, ii, iii, iv, v and vi

9.8f   **The California sea lion (*Zalophus californianus*) in Fig. 9.1 is pushing a plastic crate around a pool in a zoo. The crate contains plastic balls and small dead fish that fall through the gaps between the balls and into the water when the crate is moved. This device is a type of**

a. environmental enrichment

b. immersion device

c. ecological enrichment

d. environmental enhancement

Fig. 9.1.

9.9f **Which of the following statements about stereotypic behaviour is false?**

a. It is sometimes reversible

b. Many scientists believe that it is an indicator of poor welfare

c. There is no evidence of physical changes to the brain in animals that exhibit stereotypic behaviour

d. Stereotypic behaviour is not exclusively seen in zoos

9.10f **The five freedoms were established by the *Report of the Technical Committee to Enquire into the Welfare of Animals kept***

a. *in Zoological Gardens*

b. *under Human Care*

c. *in Laboratories*

d. *under Intensive Livestock Husbandry Systems*

9.11f **Which animals are kept in captivity for their ability to produce bile for the traditional Asian medicine industry?**

a. Tigers

b. Pangolins

c. Snakes

d. Bears

9.12f **Which of the following statements about animal suffering is false?**

a. We tend to interpret the behavioural and functional states of animals with reference to human experience

b. Historically, animal welfare has been focussed on reducing or eliminating negative experiences

c. Experiencing hunger is always an experience that leads to suffering

d. There is no single entity that defines suffering

9.13f **Which of the following dog breeds has the highest risk of hip dysplasia?**

a. Doberman

b. German shepherd

c. Jack Russell terrier

d. Rottweiler

9.14f **Mastitis is a disease that affects a**

a. horse's hoof

b. pig's snout

c. goat's jaw

d. cow's udder

9.15f **Which of the following has/have been removed in a polled animal?**

a. tail

b. horns

c. testes

d. ovaries

9.16f **A state of anxiety that an animal is unable to control is known as**

a. distress

b. stress

c. tension

d. pressure

9.17f **The use of which type of animal trap is banned in the Member States of the European Union and many other countries?**

a. Leghold trap

b. Sherman trap

c. Harp trap

d. Mist net

9.18f **Which of the following taxa have historically been hunted with dogs for sport in the United Kingdom?**

i. Otters

ii. Foxes

iii. Deer

iv. Hares

v. Wild boar

a. ii and iii

b. i, ii and iv

c. i, ii, iii and v

d. i, ii, iii, iv and v

9.19f The 'five freedoms' were first proposed in the

a. 1940s

b. 1950s

c. 1960s

d. 1970s

9.20f Complete the following sentence using the terms below: 'In 1995 B. W. Boat published a paper on the relationship between violence to ………… and violence to ………….'

a. cats/dogs

b. wild animals/companion animals

c. children/animals

d. women/cats

Intermediate

9.1i 'Canned hunting' is the hunting of wild animals for 'sport' in a fenced off area from which they cannot escape. In South Africa this activity involves mostly

a. African buffalo (*Syncerus caffer*)

b. lions (*Panthera leo*)

c. greater kudu (*Tragelaphus strepsiceros*)

d. gemsbok (*Oryx gazella*)

9.2i Brachycephalic airway obstruction syndrome is associated with which of the following dog breeds?

a. Corgi

b. Pomeranian

c. Yorkshire terrier

d. Pug

9.3i The removal of skin from the hindquarters of some breeds of sheep is called mulesing and is often carried out without adequate pain relief. It is done to

a. reduce the risk of flystrike

b. prevent dystocia

c. prevent scabies

d. prevent dermatitis

9.4i Piling is behaviour that may result in suffocation and death in

a. chickens

b. pigs

c. calves

d. goats

9.5i Pinioning is a method of restraint that involves mutilation and is used to prevent the escape of

a. horses

b. pigs

c. cattle

d. birds

9.6i Aphagia is a condition in which an animal is unable to

a. sleep

b. swallow

c. walk

d. reproduce

9.7i   **Non-ambulatory disabled cattle are unable to**

    a. reproduce

    b. successfully rear their young

    c. rise from a recumbent position or walk

    d. raise their head

9.8i   **Small cages for laying hens that are stacked in rows and on top of each other are called**

    a. stacked cages

    b. battery cages

    c. sequential cages

    d. vertical cages

9.9i   **A farrowing crate is a device used to contain a sow when she has piglets. It is specifically designed to**

    a. prevent her from feeding continuously

    b. prevent her from turning around and crushing her piglets

    c. allow easy access for a vet

    d. prevent boars from accessing her piglets

9.10i   **The framework used by many zoos to maintain a successful animal enrichment programme has the acronym**

    a. BEETLE

    b. WASPS

    c. SPIDER

    d. BEES

9.11i   **Which of the following have been most important in greatly increasing lamb survival in sheep flocks in Australia, New Zealand and the United Kingdom?**

    a. Housing improvements

    b. Changes in lambing practices

    c. Improvements in stock handling practices

    d. Vaccines and anthelmintics

9.12i In which of the following countries has the red fox (*Vulpes vulpes*) been deliberately introduced – i.e. released where it does not occur naturally – to provide animals for fox hunting?

    a. Germany

    b. New Zealand

    c. Poland

    d. Australia

9.13i A captive bolt is used

    a. to humanely slaughter livestock

    b. as an an enrichment device in zoos

    c. to temporarily tether some farm animals

    d. during surgery

9.14i Behavioural problems in some animals can be treated by repeatedly pairing a stimulus that elicits an unpleasant response with something that is emotionally positive until a positive association is made between the stimulus and the new response. This is known as

    a. habituation

    b. flooding

    c. counterconditioning

    d. overlearning

9.15i Narayan *et al.* (2013) found that when male koalas (*Phascolarctos cinereus*) (Fig. 9.2) were handled in zoos they had 200% higher FCM levels compared with non-handled males. FCM is a measure of

    a. bacterial contamination

    b. stress

    c. heart rate

    d. respiratory rate

Fig. 9.2.

9.16i **Slat-chewing is an abnormal behaviour commonly observed in**

    a. chickens

    b. sheep

    c. farmed salmon

    d. turkeys

9.17i **The administration of drugs to animals (e.g. mice) in scientific studies via a tube into the stomach and the force feeding of geese and ducks are both achieved by using**

    a. a gavage tube

    b. an intravenous line

    c. an intramuscular tube

    d. a fistula

9.18i **Complete the following sentence with the name of one of the animals listed below: Many major grocery chains no longer sell certain brands of coconut milk because the coconut farmers in Thailand use inhumane methods to train .... to pick coconuts.**

    a. Sumatran orangutans (*Pongo abelii*)

    b. lar gibbons (*Hylobates lar*)

c. siamangs (*Symphalangus syndactylus*)

d. pig-tailed macaques (*Macaca nemestrina*)

9.19i Lameness in dairy cows is directly influenced by

a. feed type

b. the behaviour of stockmen/stockwomen

c. the rate of parasitic infection

d. the weather

9.20i In 1935 the number of domestic cats collected by the RSPCA in Liverpool, England, was 37,352 but by 1942 it had fallen to 18,300 (Rees, 1982). By 1945 the total had increased again to 29,500. The most likely reason for this increase was

a. the social deprivation and loss of housing stock caused by bombing raids during World War II

b. uncontrolled breeding among feral cat populations

c. a discontinuation of measures to control cats in the city

d. increased affluence in the city resulting in increased pet ownership

Advanced

9.1a Which of the following milking systems offers the most welfare benefits for cattle?

a. A guided robotic milking system

b. A free-flow robotic milking system

c. A carousel milking system

d. A herringbone milking system

9.2a The Edinburgh Family Pen System is a housing system used on some farms to improve the welfare of

a. chickens

b. sheep

c. pigs

d. goats

9.3a The *RSPCA Welfare Standards for Farmed Atlantic Salmon* state that in an emergency a sick or badly injured salmon may be killed using 'a priest of appropriate size for the fish'. A priest is

a. a device for injecting a lethal dose of an anaesthetic

b. a type of captive bolt

c. a device used for depriving the fish of oxygen

d. a small mallet

9.4a How many inherited disorders are known in dogs?

a. >300

b. >500

c. >700

d. >1000

9.5a Which of the following cause(s) cat flu?

i. Feline herpes virus 1 (FHV1)

ii. Feline leukaemia virus (FeLV)

iii. Feline calicivirus (FCV)

iv. Feline infectious peritonitis

a. i

b. i and ii

c. i and iii

d. ii, iii and iv

9.6a The use of chemical or mechanical means to cause pain to the front feet and legs of horses when they place them on the ground in order to alter their gait for aesthetic reasons – so they lift their feet higher than normal – is known as

a. soring

b. springing

c. blistering

d. scorching

9.7a **The Society for the Protection of Animals Liable to Vivisection was founded in 1875 by**

a. Jeremy Bentham

b. Frances Cobb

c. Richard Ryder

d. Richard Martin

9.8a **The Hurnik-Morris housing system is designed to provide improved welfare standards – including opportunities for socially co-ordinated eating and resting, physical exercise and access to the opposite sex – for**

a. turkeys

b. chickens

c. goats

d. pigs

9.9a. **The number of individuals that develop a specific illness or condition over a specific period of time is the**

a. morbidity rate

b. mortality rate

c. natality rate

d. secondary attack rate

9.10a **Leghold traps are made in which of the following variants?**

i. Padded

ii. Offset

iii. Laminated

a. i and ii

b. i and iii

c. ii and iii

d. i, ii and iii

9.11a Compared with rats and mice kept in barren conditions, those kept in enriched conditions have

i.   heavier brains

ii. increased neural branching

iii. increased synapse densities

iv. better memories

v.   better learning abilities

vi. better problem-solving abilities

a. i, ii and iii

b. iv, v an vi

c. i, ii, iv and vi

d. i, ii, iii, iv, v and vi

9.12a Following more than three decades of protests and campaigning by PETA, which of the following events occurred?

a. In 2017 Ringling Bros. and Barnum and Bailey Circus announced it was closing down after 146 years of operation

b. The United Kingdom Government banned hunting with dogs

c. A moratorium on commercial whaling was introduced by the International Whaling Commission

d. The closure of all roadside zoos in the United States

9.13a *Kaika*, *Ham*, *Félicette* and *Albert II* were the names of animals that were

a. taught American Sign Language

b. sent into space

c. rescued from roadside zoos in the United States by PETA

d. chimpanzees sent to a sanctuary after living for many years at a drug testing facility

9.14a **The 'forced swim test' involves putting a small animal (e.g. a mouse or rat) into a vertical tube filled with water from which there is no escape and measuring how long it continues to swim and attempt to climb out. It is used to measure the efficacy of potential treatments for**

a. schizophrenia

b. obsessive-compulsive disorder

c. depression

d. attention deficit hyperactivity disorder

9.15a **The hormone that stabilises mood and feelings of wellbeing is called**

a. serotonin

b. relaxin

c. thyroxin

d. melatonin

9.16a **In the European Union, the Laying Hens Directive requires which of the following to be available to hens reared in enriched cages?**

i. A minimum floor area per bird

ii. A nest

iii. Litter that allows pecking and scratching

iv. Perches

v. A drinking system

vi. A feed trough available without restriction

vii. Claw-shortening devices

a. i, ii, iii, v and vi

b. ii, iii, v, vi and vii

c. i, ii, iii, iv, v and vi

d. All of the items listed are required by the Directive

9.17a Table 9.2 shows the animals taken in by the 152 centres belonging to the Royal Society for the Prevention of Cruelty to Animals (RSPCA) in England and Wales in 2019 (RSPCA, 2019). Match the data in columns A–D with the correct animal types

Table 9.2

	A	B	C	D
Dogs	29,432	10,564	17,321	17,321
Cats	10,564	29,432	29,432	29,432
Rabbits	3,180	3,180	10,564	3,180
Wild animals	17,321	17,321	3,180	10,564

   a. A

   b. B

   c. C

   d. D

9.18a A study of 261 prison inmates in the United States (Hensley and Tallichet, 2005) found that

   a. those who experienced animal cruelty at a younger age were more likely to demonstrate recurrent animal cruelty themselves

   b. those who observed a friend abusing animals were more likely to hurt or kill animals more frequently

   c. those who were young when they first witnessed animal cruelty hurt or killed animals themselves at a younger age

   d. all of the above were true

9.19a Alpha-mannosidosis causes ataxia in calves and is caused by a deficiency of the enzyme lysosomal α-mannosidase. The disease is caused by a recessive allele. Homozygotes have very little of the enzyme; heterozygotes have about half of the normal enzyme activity and are otherwise normal. If two heterozygotes breed, what is the probability that one of their offspring will inherit the disease?

   a. zero

   b. 0.25

c. 0.50

d. 0.75

9.20a Films made in the United States that use live animals include *No Animals Were Harmed* end credit certification operated by

a. American Humane (American Humane Society)

b. People for the Ethical Treatment of Animals (PETA)

c. The American Society for the Prevention of Cruelty to Animals (ASPCA)

d. The Humane Society of the United States (HSUS)

10 Animal Rights, Ethics and Law

This chapter contains questions about human attitudes towards animals, legal personality, the development of animal rights, animal welfare organisations, and the legal protection of animals exploited by humans.

Foundation

10.1f The study of animal rights is a branch of

    a. biology

    b. philosophy

    c. sociology

    d. psychology

10.2f Aristotle believed that humans were superior to animals because they could

    a. count

    b. speak

    c. reason

    d. walk on two legs

10.3f A stimulus that causes pain, fear or emotional discomfort is called

    a. an aversive stimulus

    b. an immersive stimulus

© Paul A. Rees 2022. *Key Questions in Animal Behaviour and Welfare: A Study and Revision Guide* (P.A. Rees)
DOI: 10.1079/9781789248975.0010

c. an adversarial stimulus

d. an emotive stimulus

10.4f Someone who does not eat meat or other animal products, generally for ethical, health or environmental reasons is known as a

a. vegetarian

b. vegan

c. lacto-vegetarian

d. pescatarian

10.5f 'The capacity to have positive and negative experiences, usually thought of as happiness and suffering' is one definition of

a. capability

b. consciousness

c. sentience

d. awareness

10.6f The Australian philosopher Peter Singer is credited with founding the modern philosophical basis for the establishment of the animal liberation movement in 1975 with the publication of his book

a. *Animal Rights*

b. *Animal Cruelty*

c. *Animal Liberation*

d. *Animal Ethics*

10.7f The study of the interactions between humans and other animals is called

a. anthropology

b. anthrozoology

c. sociobiology

d. biological anthropology

10.8f In many cultures using animals for entertainment and gambling on the outcome of fights between them is common. Camel wrestling is most commonly associated with

a. Morocco

b. Mongolia

c. Egypt

d. Turkey

10.9f Which of the following is not an acronym or abbreviation for an organisation that is dedicated to reducing animal suffering in captivity?

a. HSI

b. CAPS

c. ADI

d. IUCN

10.10f The only civil rights organisation in the United States that works to achieve legal rights for species other than *Homo sapiens* is called the

a. Nonhuman Rights Project

b. Sentient Beings Project

c. Primate Rights Project

d. Animal Rights Project

10.11f Which moral philosopher said, in relation to our obligation to treat certain animals with compassion, 'The question is not, can they reason? nor, can they talk? but, can they suffer?'

a. Immanuel Kant

b. Jeremy Bentham

c. Ludwig Wittgenstein

d. Peter Singer

10.12f PETA is an acronym for

a. Provide Ethical Treatment for Animals

b. People for the Equitable Treatment of Animals

c. People for the Ethical Treatment of Animals

d. Preventing Experimental Trials on Animals

10.13f Which of the following has not published extensively on the subject of environmental enrichment in zoos?

a. Robert Young

b. David Shepherdson

c. Vernon Kisling

d. Hal Markowitch

10.14f Animals used in vivisection are always

a. alive

b. mammals

c. dead

d. unable to feel pain

10.15f Anthropomorphism is the tendency to attribute to animals

a. human characteristics

b. childlike characteristics

c. aggressive characteristics

d. behaviours based on their shape

10.16f Which of the following characteristics of animals was not considered by Singer as important in determining whether or not they should be afforded rights?

a. The ability to plan for the future

b. Bipedalism

c. A complex social life

d. The ability to communicate

10.17f In determining which animals are sentient it has been argued that we should not require absolute certainty before any particular species is afforded a degree of protection under the law. This principle is known as the

a. uncertainty principle

b. confidence principle

c. ambiguity principle

d. precautionary principle

10.18f Which of the following cannot be a legal person?

i.   A corporation

ii. The City of New York

iii. Part of the Amazon rainforest in Colombia

iv. The Whanganui River, New Zealand

v. A rhesus macaque (*Macaca mulatta*)

a. i, ii, iv and v

b. iii, iv and v

c. v

d. iv and v

10.19f In legal terms the words 'person' and 'human' are

a. not synonyms

b. mutually exclusive

c. not antonyms

d. similes

10.20f Cockfighting and dog fighting were banned in England and Wales in

a. 1801

b. 1835

c. 1859

d. 1872

Intermediate

10.1i **In the 13ᵗʰ century, who argued that people should be kind to animals so that they did not pick up cruel habits and, as a result, treat other people badly?**

a. St Thomas Aquinas

b. St Benedict II

c. St Hilary

d. St Francis of Assisi

10.2i **PETA was formed in**

a. 1960

b. 1970

c. 1980

d. 1990

10.3i **The Oxford Centre for Animal Ethics was founded in 2006 by**

a. Richard Ryder

b. Tom Regan

c. Mary Midgley

d. Andrew Linzey

10.4i **'Speciesism' is the view that discriminating against animals is wrong and the animal equivalent of racism in humans. The term was first coined by**

a. Peter Singer

b. Richard Ryder

c. Tom Regan

d. Jeremy Bentham

10.5i **Animals usually have no standing in a court of law because they do not possess**

a. a legal personality

b. a legal character

c. an authorised temperament

d. a moral character

10.6i **The ancient Indian principle of nonviolence to all living things that is a key virtue in Hinduism, Buddhism and Jainism is known as**

a. satya

b. asteya

c. ahimsa

d. brahmacarya

10.7i **Which was the first country in the world to pass a law to prevent cruelty to animals?**

a. United Kingdom, in 1822

b. United States, in 1815

c. France, in 1819

d. Italy, in 1826

10.8i **The oldest animal welfare organisation in the world is**

a. the American Society for the Prevention of Cruelty to Animals

b. the Royal Society for the Prevention of Cruelty to Animals

c. Deutscher Tierschutzbund

d. Swiss Animal Protection

10.9i **Which of the following animals are treated as sacred in Hinduism?**

i. Elephants

ii. Cattle

iii. Horses

iv. Tigers

v. Lions

a. ii

b. i and ii

c. i, ii and iv

d. i, ii, iii, iv and v

10.10i The rightness or wrongness of a given act may be determined by the extent to which its consequences best serve all affected by its outcomes. This philosophy is called

a. epistemology

b. utilitarianism

c. existentialism

d. phenomenology

10.11i The intrinsic value of an animal's life is determined by

a. its legal status

b. its species

c. its existence

d. its phylum

10.12i The attitude to animals whereby they are treated as resources to be bought and sold is known as

a. capitalisation

b. commodification

c. utilisation

d. exploitation

10.13i Complete the following with one of the options listed below: 'In human societies individuals consent to surrender some of their freedoms in exchange for the protection of their remaining rights. This is known as a .... Animals are incapable of entering into such arrangements'.

a. social contract

b. societal agreement

c. social concord

d. collective agreement

10.14i Match the academics interested in animal rights listed in Table 10.1 with their nationality, affiliation and expertise.

Table 10.1

	A	B	C	D
American philosopher, North Carolina State University	Richard Ryder	Peter Singer	Tom Regan	Richard Ryder
Australian philosopher, Princeton University	Tom Regan	Richard Ryder	Peter Singer	Peter Singer
English psychologist and philosopher, Tulane University	Peter Singer	Tom Regan	Richard Ryder	Tom Regan

    a. A

    b. B

    c. C

    d. D

10.15i For an animal to possess autonoetic consciousness it must be capable of

    a. mental time travel

    b. counting

    c. self-recognition

    d. forming social relationships

10.16i The toxicity of chemicals to humans is widely measured using animal models. The test used to determine the sensitivity of the eye or skin to cosmetics, pharmaceuticals and other products using animals is called the

    a. Fraize test

    b. Praize test

c. Draize test

d. Traize test

10.17i The guiding principle applied to animal testing known as the 'Three Rs' refers to

a. replacement, reduction and release

b. redundancy, refinement and replacement

c. refinement, reassignment and replacement

d. replacement, reduction and refinement

10.18i Immanuel Kant believed that a 'being' has value and is thus morally considerable – and a 'thing' does not – because a being possesses the quality of

a. intelligence

b. personality

c. sentience

d. personhood

10.19i The view that it is acceptable to keep animals in zoos if it can be shown that they are able to experience a quality of life that is at least as good as that experienced by their conspecifics in the wild is the

a. analogous life test

b. equivalent life test

c. comparable life test

d. similar life test

10.20i The name given to a series of events following an illegal vivisection performed by the surgeon William Bayliss on a dog at the Department of Physiology of University College, London in 1903 in front of medical students was

a. the White Dog Affair

b. the Brown Dog Affair

c. the Spotted Dog Affair

d. the Black Dog Affair

Advanced

10.1a Who wrote *Animals and Why They Matter*?

    a. Andrew Linzey

    b. Tom Regan

    c. Peter Singer

    d. Mary Midgley

10.2a Which of the following taxa have been scientifically shown to be insentient?

    a. Fishes

    b. Crustaceans

    c. Cephalopods

    d. None of the above

10.3a Many people love their pet cat or dog but still eat meat. The anxiety caused by this mismatch between feelings and behaviour is known as

    a. cognitive dissonance

    b. cognitive discord

    c. emotional conflict

    d. cognitive divergence

10.4a *Habeus corpus* lawsuits have been used in some law courts on behalf of chimpanzees (*Pan troglodytes*) in attempts to have them released from captivity. A writ of *habeus corpus* is normally used to examine whether or not

    a. an animal is being cruelly treated

    b. an animal should be given human rights

    c. the detention of a human against his or her will is lawful

    d. an animal should be treated as possessing a legal personality

10.5a Who published a book entitled *The Case for Animal Rights* in 1983?

    a. Donald Griffin

    b. Tom Regan

    c. Ruth Harrison

    d. Peter Singer

10.6a In the case of _The Nonhuman Rights Project on Behalf of Tommy v. Patrick Lavery_, filed in Fulton County Court, New York, who was Tommy?

    a. An Asian elephant (_Elephas maximus_)

    b. A gorilla (_Gorilla gorilla_)

    c. A dolphin (_Tursiops truncatus_)

    d. A chimpanzee (_Pan troglodytes_)

10.7a Complete the following sentence with one of the options below: 'In a lawsuit against SeaWorld, PETA argued that …………… were being held as slaves in violation of the 13th Amendment to the US Constitution'.

    a. five wild-caught orcas

    b. five captive-bred orcas

    c. five wild-caught dolphins

    d. five captive-bred dolphins

10.8a In the legal case cited in Q10.7a it was argued that the wording of the law does not require the defendant to

    a. be a person

    b. be a US citizen

    c. live in the United States

    d. have _locus standi_

10.9a Consideration of the moral importance of each individual animal and especially the 'maximum sufferer' is emphasised in the theory of

    a. mind

    b. painism

    c. speciesism

    d. sentientism

10.10a **The movement that advocates that we should help in which-ever causes (including animal rights) allow us to do most good is the**

  a. effective egotist movement

  b. ineffective egotist movement

  c. effective altruist movement

  d. ineffective altruist movement

10.11a **Which was the first country to formally protect the 'dignity' of animals in its constitution?**

  a. United Kingdom

  b. United States of America

  c. Switzerland

  d. Austria

10.12a **Which country granted non-human hominids certain rights under the law in 1999?**

  a. Australia

  b. United States

  c. Germany

  d. New Zealand

10.13a **Some philosophers argue that humans have moral status and non-human animals do not because there are distinctly human capacities that only humans possess. This viewpoint is known as**

  a. human exceptionalism

  b. antispeciesism

  c. speciesism

  d. existentialism

10.14a **A serious problem with the use of the concept of personhood in determining who should be afforded moral consideration is that it excludes some humans such as infants and the severely cognitively disabled. In the literature on animal rights these cases are known as**

a. exceptional cases

b. marginal cases

c. trivial cases

d. extraordinary cases

10.15a Complete the following sentence using the name of one of the countries listed below. 'In a court decision of 2015 an orangutan (*Pongo* sp.) called *Sandra* was awarded nonhuman personhood rights by a court in ....' and sent to live at the Center for Great Apes in Wauchula, Florida.

a. Argentina

b. Mexico

c. Brazil

d. Costa Rica

10.16a In some jurisdictions groups of individuals have attempted to bring cases to the courts on behalf of nonhuman animals because the animals would not have been recognised as plaintiffs themselves. In 1988 a German court rejected a lawsuit brought by a group of ecologists on behalf of

a. seabirds affected by pollution in the North Sea

b. seals affected by pollution in the North Sea

c. fish poisoned by toxins in the River Rhine

d. dolphins caught in fishing nets in the Mediterranean Sea

10.17a BORN FREE USA. *et al.*, Plaintiffs, v. GALE NORTON, Secretary, Department of the Interior *et al.*, Defendants, and THE ZOOLOGICAL SOCIETY OF SAN DIEGO, *et al.*, Intervenor-Defendants was a case brought before the United States District Court for the District of Columbia in 2003 in an attempt to

a. effect the release of a female gorilla from the San Diego Zoo to an ape sanctuary

b. prevent the importation of wild African elephants from Swaziland to zoos in the United States

    c. force the closure of the San Diego Zoo

    d. effect the release of a family of seven chimpanzees from the San Diego Zoo to a chimpanzee sanctuary

10.18a Some companion and farm animals have been given some rights by virtue of the imposing of minimum dimensions for their accommodation. The government of which of the countries listed below published a *Code of Practice for the Welfare of Rabbits* in 2009 that indicated that the minimum size of accommodation for a rabbit (*Oryctolagus cuniculus*) should allow it to hop three times from one end of the shelter to the other?

    a. Canada

    b. New Zealand

    c. Wales

    d. Switzerland

10.19a The first animal cruelty law in the United Kingdom was an Act to Prevent the Cruel and Improper Treatment of Cattle 1822. The term 'cattle' referred to

    a. cattle only

    b. cattle and horses only

    c. cattle, horses, mules, asses and sheep only

    d. cattle, horses, mules, asses, sheep and similar livestock

10.20a Complete the following sentence using one of the terms listed below: 'In 1996, lawyers attempted to bring a case at the Mito District Court in Japan on behalf of a population of .... to prevent damage to their habitat by development: a marshy area on the shores of Lake Kasumigaura'.

    a. cranes

    b. ducks

    c. geese

    d. swans

11 Answers

A multiple choice question has a stem (the 'question'), a key (the 'answer') and a number of distracters (wrong answers intended to distract the student from the key). This part of the book contains the key to each question along with a brief explanation of why this is correct and, in some cases, what the distracters mean.

Chapter 1

1.1f	C	Ethology is the study of the behaviour of animals in their natural environment rather than in laboratory conditions.
1.2f	B	The Austrian scientist Konrad Lorenz and the Dutch scientist Nikolaas Tinbergen were largely responsible for founding ethology.
1.3f	D	The concept of a black box in this context is analogous to that of a black box as a safety device in an aircraft. Data enters the black box and data leaves it; a knowledge of how this data is processed within the box is not important to understanding the functioning of the aircraft. The distracters are fictitious in this context.
1.4f	A	Clutton-Brock studied red deer in Scotland and meerkats in South Africa.
1.5f	D	The behaviour of an animal is a product of the genes it has inherited and its interactions with its physical environment and with other organisms of the same and different species.
1.6f	B	Darwin's *On the Origin of Species* considered the possibility that the behavioural traits of animals could be inherited, although at the time the mechanism of inheritance was not understood. The distracters are other biologists who made important contributions to the study of taxonomy and evolution.

© Paul A. Rees 2022. *Key Questions in Animal Behaviour and Welfare: A Study and Revision Guide* (P.A. Rees)
DOI: 10.1079/9781789248975.0011

1.7f	A	Behaviourism is correct. The distracters are other schools of psychology.
1.8f	B	Sociobiology attempts to explain social behaviour in terms of evolution.
1.9f	A	Zuckerman was a zoologist who began his career working for the Zoological Society of London.
1.10f	D	These books were written by the zoologist Dr Desmond Morris who was Curator of Mammals at London Zoo in the early part of his career. In these books Morris wrote about humans from the point of view of a zoologist.
1.11f	B	Comparative psychology is the branch of psychology which investigates the behaviour of different types of animals in an attempt to understand human behaviour.
1.12f	A	The English philosopher Jeremy Bentham wrote this book. The distracters are other philosophers and writers on animal rights.
1.13f	D	Animal cruelty laws cover all of these types of animal cruelty.
1.14f	D	All of these journals, and many others, publish studies of the behaviour of animals in zoos.
1.15f	C	Pavlov was a Russian physiologist who worked extensively on the digestive system.
1.16f	C	The Edward Grey Institute of Field Ornithology is part of the Department of Zoology of the University of Oxford.
1.17f	B	Welfare is concerned with the extent to which an animal is able to cope with its environment.
1.18f	C	*Blackfish* is a film about the orca *Tilikum* who performed at SeaWorld (Orlando, Florida) and the consequences of keeping killer whales in captivity.
1.19f	A	An EEG records brain activity. The first EEG recording of a human brain was made in 1924.
1.20f	D	Lack and Neal were both schoolteachers. Lack eventually became Director of the Edward Grey Institute at the University of Oxford.
1.1i	D	Piaget was a Swiss psychologist who worked on the cognitive development of children.
1.2i	B	Lorenz, Tinbergen and von Frisch shared the Nobel Prize for their work in ethology.
1.3i	C	This study was initiated by Dr Jane Goodall. John Napier and Dian Fossey were both primatologists. Robert Hinde supervised the PhD studies of Jane Goodall and Dian Fossey.
1.4i	B	Prof. F.W.R. Brambell led this committee. The distracters are fictitious in this context.
1.5i	A	Cynthia Moss has made a long-term study of the elephants of Amboseli.

1.6i	C	The first animal behaviour journal was published in Germany in 1937: *Zeitschrift für Tierpsychologie*. It was renamed *Ethology* in 1986.
1.7i	B	John Wray (Ray) published a book on instinct in birds.
1.8i	D	Harlow worked on the effect of maternal deprivation on the behaviour and development of monkeys.
1.9i	D	Skinner was a behaviourist not an ethologist.
1.10i	A	Tinbergen worked extensively on gulls and wrote *The Herring Gull's World*, published in 1953.
1.11i	B	Behaviourists focussed on the role of learning in animal behaviour.
1.12i	B	*Mental Evolution in Animals* was published in 1883; *Animal Machines* was published in 1964; *The Question of Animal Awareness* was published in 1976; *The Case for Animal Rights* was published in 1983.
1.13i	C	*In the Shadow of Man* was written by Dr Jane Goodall and tells the story of her early work with the chimpanzees in Gombi.
1.14i	A	A long-term selective breeding experiment examined the role of inheritance in silver foxes. It was conducted at the Institute of Cytology and Genetics in Siberia.
1.15i	A	This paper was written by Julian Huxley. The distracters are other individuals who were interested in birds and/or behaviour.
1.16i	B	Most people who participated in this survey kept animals primarily for companionship.
1.17i	D	Köhler wrote *The Mentality of Apes*. The other titles are fictitious.
1.18i	B	Bowlby studied development in children drawing on the work done by Harlow on attachment in monkeys.
1.19i	B	Douglas-Hamilton studied the elephants of Lake Manyara National Park while working for his PhD at the University of Oxford.
1.20i	D	Hal Markowitz was a pioneer in environmental engineering and behavioural enrichment for animals living in zoos.
1.1a	B	Ruth Harrison wrote *Animal Machines*. The other titles are fictitious in this context.
1.2a	A	In 1986 Prof. Donald Broom was appointed the first Professor of Animal Welfare in the world.
1.3a	B	Broom was Professor of Animal Welfare in the Department of Veterinary Medicine, University of Cambridge.
1.4a	C	*On Aggression* was written by Konrad Lorenz. The distracters are other authors of books concerned with behaviour.
1.5a	A	The Max Plank Institute of Animal Behaviour was once the workplace of Konrad Lorenz. The distracters are fictitious.
1.6a	D	The German zoologist Ernst Haeckel introduced the term 'ecology'. An ethologist was a term for an actor or mimic in the 17th century. In the 18th century 'ethology' was the science of ethics.

1.7a	B	Schaller wrote *The Serengeti Lion*. The distracters are other well-known field biologists.
1.8a	D	Tinbergen founded this group after his arrival at Oxford in the 1940s.
1.9a	A	The first animal cruelty law passed in the United Kingdom protected farm mammals.
1.10a	B	The term 'evolutionarily stable strategy' was first used by John Maynard Smith.
1.11a	C	These questions are known as 'Tinbergen's four questions'.
1.12a	A	Behavioural ecologists are also known as adaptationists because they study the ways in which ecological pressures result in the evolution of adaptive behaviour.
1.13a	C	Edward O. Wilson was the founder of the science of sociobiology and published his ideas in a book entitled *Sociobiology: The New Synthesis* in 1975.
1.14a	B	Behavioural ecology is the study of the behavioural adaptations that animals make in response to selective pressures in the environment; the effect of evolution on behaviour.
1.15a	D	Robert Trivers is an American evolutionary biologist.
1.16a	B	This centre is named after the famous primatologist Robert Yerkes. The distracters are other famous primatologists.
1.17a	C	John Maynard Smith wrote this book. He originally trained as an aeronautical engineer and later became a geneticist. He went on to develop important concepts in theoretical and mathematical ecology while at the University of Sussex. The distracters are other scientists with an interest in mathematical biology.
1.18a	A	The first veterinary school was established in 1762 in Lyon, France.
1.19a	A	This paper was written by the American psychologist Harry Harlow.
1.20a	D	Gestalt psychology considers the elements of mental life as a whole.

Chapter 2

2.1f	A	Positive phototaxis is a movement toward light.
2.2f	B	Habituation is correct. It is adaptive for individuals to stop responding to stimuli that have no survival implications for them.
2.3f	A	This is inclusive fitness and consists of direct fitness (from one's own young) + indirect fitness (from providing assistance to relatives).
2.4f	C	Neophobia is a fear of new things.
2.5f	D	Cephalisation is the concentration of sense organs at one end (the head end) of an organism. This allows the animal to obtain stimuli from a new environment as soon it enters it.

2.6f	A	When the model is moved from left to right (A) it appears like the silhouette of a bird of prey: a bird with a short neck and a long tail. When it moves in the opposite direction it looks like that of a duck or goose: a bird with a long neck and a short tail.
2.7f	B	A positive geotaxis is a movement downwards, in the direction of the force of gravity.
2.8f	D	The pondweed dance is fictitious.
2.9f	A	The honey guide and the bees are mutually dependent in a symbiotic relationship.
2.10f	B	Kainism (or siblicide) is the killing of siblings. This occurs, for example, when an older nestling kills a younger one.
2.11f	A	Myrmecophagous species eat ants.
2.12f	B	This behaviour has evolved independently in different taxa so is an example of convergent evolution.
2.13f	C	Direct fitness relates to the direct transmission of genes to the next generation by reproduction.
2.14f	B	A superstimulus is an exaggerated stimulus that is similar to the normal stimulus to which there is an existing tendency to respond, e.g. male sticklebacks will attack a wooden model fish with a bright red underside more than a real male if the model is redder.
2.15f	A	The interpretation of the stimuli received by an organism is called perception.
2.16f	C	This is a kinesis. In klinotaxis there is an alteration in the frequency of turning in the animal's random movements. In orthokinesis the rate of movement alters.
2.17f	D	A matutinal species is active at dawn or in the early morning, for example bees.
2.18f	C	Aestivation is a period of inactivity and seclusion used to avoid periods of high temperatures.
2.19f	A	Some insects exhibit flower loyalty (or flower constancy), i.e. they visit only one flower type or species.
2.20f	D	Sensory input is transmitted from the sensory receptors via the peripheral nervous system to the central nervous system. Here it is integrated and then the motor output is transmitted to the effector organs.
2.1i	B	Some harmless insects possess morphological features that are designed to intimidate their potential predators, e.g. moths, with patterns resembling large eyes on their wings?
2.2i	B	This sequence is called a reaction chain.
2.3i	A	This is a fixed action pattern and it is initiated after exposure to a releaser.

2.4i	D	Baird's tapir does not have a harem mating system. In the other three species a single male mates with a group of females.
2.5i	C	Many species of animals possess red structures that function as sign stimuli, e.g. the red belly of male sticklebacks, the red breast of European robins, the red patch on the bill of a herring gull.
2.6i	A	This is a definition of cognition: what animals know and how they know it.
2.7i	B	The term 'meme' was first used by Richard Dawkins in his book *The Selfish Gene* in 1976. A deme is a group of randomly mating organisms.
2.8i	C	Chemotaxis is a directional response to a chemical gradient.
2.9i	B	This is intraspecific brood parasitism because the host and parasite are the same species.
2.10i	C	Autotomy is the self-amputation of a body part.
2.11i	D	Male bowerbirds construct elaborate structures to attract females.
2.12i	A	Mass spawning is a type of reproductive synchrony whereby both sexes release their gametes into the environment at the same time.
2.13i	A	Kleptoparasitism is a behaviour whereby an animal obtains food by stealing it from another.
2.14i	D	Releaser is another name for a sign stimulus. The distracters are fictitious in this context.
2.15i	B	Birds are only able to begin to fly when their nervous systems and muscles reach a certain stage of maturity.
2.16i	C	Social grooming is important in many primate societies because it helps to maintain dominance hierarchies, low-ranked individuals generally grooming high-ranked individuals.
2.17i	C	Phenotypic matching is a method of recognising relatives. There is evidence that rhesus macaques (*Macaca mulatta*) are capable of visual phenotypic matching and recognise relatives from similarities in their faces.
2.18i	D	This behaviour has become ritualised and, in the context of courtship, has lost its original purpose.
2.19i	D	This is a visual cliff. Individuals of some species placed on surface A will not venture onto the glass because they perceive this as a 'cliff edge' when they see the patterned surface below (B).
2.20i	A	Mandarin ducklings will walk to the edge of surface A and jump when they reach the edge of the glass. This is because they normally hatch in a nest hole in a tree and to take their first flight they must jump from the hole.
2.1a	A	Occam's razor favours the explanation that makes the fewest assumptions. The distracters are the names of other 'principles' that are irrelevant in this context.

2.2a	B	Behavioural plasticity is the capacity of an animal to alter its behaviour in response to a change in the environment.
2.3a	C	C is correct. The distracters refer to other patterns of activity: matutinal means active at dawn or early morning; crepuscular means active at twilight (dawn and dusk); vespertine means active at dusk or early evening.
2.4a	D	This is displacement behaviour, e.g. an animal may groom when faced with a conflict between approaching or avoiding another animal.
2.5a	A	'Umwelt' is German for 'environment' or 'surroundings'; more specifically 'self-centred world'. Organisms may share the same environment but have different umwelten. For example, different species perceive different frequencies of sound. 'Verhalten' means behaviour.
2.6a	B	Richard Dawkins is a zoologist who has written extensively about evolution, especially the importance of reciprocal altruism in social species. The distracters are other scientists who have written about evolution.
2.7a	C	The red spot on the parent's beak is a sign stimulus (releaser) and the pecking behaviour of the chick is a fixed action pattern.
2.8a	B	The Mauthner neurons are located in the hindbrain and are responsible for a rapid escape reflex initiated by an acute tactile, acoustic or visual stimulus.
2.9a	A	Ovicide is correct. Siblicide (or kainism) is the killing of a sibling; patricide is the killing of one's father; matricide is the killing of one's mother.
2.10a	D	Diapause can affect all of these stages of development.
2.11a	D	An ultradian rhythm occurs over less than 24 hours; an infradian rhythm occurs over more than 24 hours; a circadian rhythm has a period of approximately 24 hours; and a circannual rhythm has a period of approximately one year.
2.12a	B	This is a feeding method used by frogs and toads whereby they rapidly extend their tongue and catch their prey with the tip.
2.13a	C	*Difflugia* is a genus of protozoans whose members are referred to as testate amoebae or shelled amoebae because of their ability to construct tests or shells of various mineral particles, depending on the environment.
2.14a	A	Dermatotrophy means 'feeding on skin'. In caecilians a nursing mother produces layers of thick skin which her offspring peel off, using specialised teeth, and eat.
2.15a	A	A fixed action pattern is an instinctive behaviour.
2.16a	B	The lizards squirt blood from their eyes by restricting blood flow from the head – thereby increasing vascular pressure – and contracting the protrusure oculi muscles of the eyelids, causing capillaries to rupture in and around the eyes.

2.17a	D	Behavioural resilience is the capacity of a particular behaviour to resist being compressed into a smaller proportion of an animal's activity budget after it has been forced to spend more time on other things, for example, searching for food.
2.18a	C	Pleiotropy is the name of the phenomenon whereby a single gene has multiple phenotypic effects on an organism.
2.19a	A	William D. Hamilton was a British evolutionary biologist who published a paper entitled *The Genetical Evolution of Social Behaviour* in 1964. This paper set out the mathematical basis of inclusive fitness. The distracters are the names of persons who published extensively on evolution.
2.20a	A	An individual's fitness is increased by supporting close relatives in rearing their young because they share many of the same genes. This contribution to overall fitness is called indirect fitness.

Chapter 3

3.1f	A	A is correct. The distracters are the same terms matched to the incorrect labels.
3.2f	D	The space between two neurons in a sequence is the synaptic cleft.
3.3f	B	The central nervous system consists of the brain and the spinal cord.
3.4f	A	A nerve impulse is the result of a reversal of the electrical charge across the membrane of the neuron.
3.5f	C	The cerebrospinal fluid has several functions including protection of the brain from physical shock.
3.6f	D	The auditory ossicles are the bones that transmit sound waves from the ear drum to the cochlea in the inner ear.
3.7f	C	The Cnidaria – jellyfishes, corals and their relatives – possess a nerve net.
3.8f	A	A lateral line is a system of sensory organs located along the head and the body of a fish (or amphibian) that detects pressure changes and movements in water.
3.9f	D	The tympanum or tympanic membrane is the ear drum.
3.10f	D	Invertebrates possess ventral solid nerve cords. Vertebrates possess a dorsal hollow nerve cord.
3.11f	B	The pit organs of pit vipers detect infrared radiation. This helps them detect 'warm-blooded' prey.
3.12f	B	The medulla oblongata is responsible for homeostatic functions such as respiration and cardiac function.
3.13f	C	Effectors are muscle or gland cells.
3.14f	D	The cerebellum is part of the vertebrate hindbrain and is concerned with the coordination of movement, posture and motor learning.

3.15f	A	Stretch receptors are mechanoreceptors in the muscles.
3.16f	C	The myelin sheath surrounds the axon in myelinated nerve fibres and is made of protein and fat.
3.17f	A	The occipital lobe is located at the back of the brain and is responsible for visual perception.
3.18f	B	B is correct. The distracters are the same terms matched with the incorrect labels.
3.19f	B	Ommatidia are the units of which the compound eyes of arthropods (e.g. insects and crustaceans) are constructed. They contain clusters of photoreceptor cells.
3.20f	D	The corpus callosum is a bundle of nerve fibres that connects the right and left hemispheres of the brain and allows communication between them.
3.1i	A	In the resting state the outside of the membrane is positively charged and the inside is negatively charged.
3.2i	C	A nerve impulse is a wave of depolarisation that travels along the axon. The outside of the axon becomes negatively charged and the inside becomes positively charged at the depolarised section.
3.3i	D	D is correct. The distracters are the same labels in incorrect orders.
3.4i	A	The Jacobson's organ detects chemicals.
3.5i	D	Chromatophores are cells that contain pigments. The distracters are other types of cells that are not involved in colour changes.
3.6i	B	B is correct. The distracters are the same labels in incorrect orders.
3.7i	C	In the 'fight or flight' response adrenaline and noradrenaline are released. The distracters are the names of other hormones.
3.8i	C	A nerve impulse 'jumps' between the sections of the axon that are not covered by the myelin sheath in a myelinated neuron.
3.9i	B	The unmyelinated section of the axon between the Schwann cells is called the node of Ranvier. The distracters are fictitious in this context.
3.10i	C	Proprioreceptors detect changes in movement and position.
3.11i	A	Echinoderms (e.g. starfish) have a nerve ring connected to radial nerves that pass along each ray (or arm).
3.12i	C	The cortex of the adrenal glands secretes cortisol.
3.13i	C	Cephalopods (octopuses and squids) have complex brains capable of solving problems.
3.14i	A	This is the refractory period.
3.15i	B	B is correct. The distracters are other parts of the brain.
3.16i	A	Synaptic vesicles contain neurotransmitters that are released into the synaptic cleft between two neurons to propagate a nerve impulse.

3.17i	C	Giant squids were used extensively to study nerve impulses because they have thick nerves that were easy to access with the tools available at the time.
3.18i	B	The section labelled 'X' is depolarised. In the resting condition the outside of the membrane is positive. When stimulated it becomes negative (depolarised).
3.19i	A	The melon is important in communication and echolocation.
3.20i	D	The utricle is a fluid-filled cavity in the inner ear that contains hair cells and otoliths (granules) that move in response to the orientation and movement of the head and send information about this to the brain.
3.1a	C	Testosterone is correct. The distracters are the names of other hormones.
3.2a	D	Stimulus filtering prevents the animal from responding to unimportant stimuli.
3.3a	B	This is the hypothalamic-pituitary-adrenal (HPA) axis.
3.4a	B	Acetylcholinesterase is an enzyme that breaks down the neurotransmitter acetylcholine after it has performed its function of transmitting an impulse from one neuron to the next. This prepares the synapse to receive the next signal.
3.5a	A	Norepinephrine (noradrenaline) functions as a hormone and a neurotransmitter.
3.6a	B	The normal functioning of the neuron depends upon the movement of sodium and potassium ions across the membrane of the axon.
3.7a	C	The organ of Corti is located in the cochlea of mammals and converts sound into electrochemical signals.
3.8a	A	Analgesics relieve pain.
3.9a	D	The hypothalamus helps to maintain homeostasis (e.g. heart rate, blood pressure, body temperature) by linking the nervous system with the endocrine system.
3.10a	B	Impulses travel faster in thick and myelinated axons than in thin or unmyelinated axons.
3.11a	C	C is correct.
3.12a	C	The semicircular canals are concerned with detecting head rotations and angular accelerations of the head.
3.13a	B	Statocysts are involved in balance and hearing.
3.14a	D	This is possible due to the horizontal separation of the eyes.
3.15a	C	C is correct. The distracters are the same colours incorrectly matched with the types of chromatophores.

3.16a	A	A small hole can act as a lens. A pinhole camera is a small box in which light passing through a small hole is focussed on a piece of photograph film inside the box to produce a negative for a photograph.
3.17a	B	These are the balance organs that coordinate flight.
3.18a	A	Prolactin is produced by the pituitary gland. It stimulates milk production and is important in maternal behaviour.
3.19a	C	C is correct. The distracters are the names of other hormones.
3.20a	D	The blind spot contains no light-sensitive cells and is the point where the optic nerve leaves the retina.

Chapter 4

4.1f	B	This is a type of associative learning called operant conditioning.
4.2f	B	An engram is a hypothetical construct that represents the physical processes and changes that represent memory.
4.3f	A	This is target training and is widely used in zoos to train animals to cooperate with veterinary examinations and procedures.
4.4f	D	A clicker, whistle or verbal praise could all act as event markers, e.g. a whistle could be blown at the time an animal performs the required behaviour prior to giving a reward.
4.5f	C	Alex was an acronym for avian language experiment/ avian learning experiment. The distracters are other bird species capable of imitating human speech.
4.6f	B	In Müllerian mimicry similar noxious species are protected by virtue of their morphological similarities. In Batesian mimicry harmless species are protected because of their morphological similarity to noxious species. In automimicry harmful (or non-toxic) members of a species are protected by virtue of their resemblance to harmful (or toxic) members of the same species. Darwinian mimicry is fictitious.
4.7f	A	Stimulus relevance is correct. For example, a rat will associate a light or a sound stimulus (but not a taste stimulus) with an electric shock consequence.
4.8f	B	This is the process of generalisation.
4.9f	D	Most of the scientific work on imprinting has been conducted on landfowl (e.g. chickens, quails etc.), ducks and geese.
4.10f	C	The main areas of the brain concerned with memory are the hippocampus, cerebellum and amygdala.
4.11f	D	Survival training for captive-bred animals is likely to involve learning to locate and select appropriate foods, avoid predators and react appropriately to the alarm signals made by conspecifics.

4.12f	B	The purpose of the outfits was to prevent young birds from imprinting on humans and habituating to their presence so they would not be attracted to people after their release.
4.13f	C	The figure illustrates classical conditioning in which the dog is conditioned to associate the sound of a whistle with food and responds by salivating.
4.14f	A	Errors will decrease with the number of trials (i.e. with practice).
4.15f	B	This is latent learning: the unconscious retention of information.
4.16f	D	In one-event learning, or single event learning, a single event results in the learning of a strong response to a stimulus, e.g. a strong fear response to pain, or aversion to a particular taste.
4.17f	C	Chance contacts with the fence cause an association to form between contact with the wire and an electric shock.
4.18f	A	This phenomenon was first described by Köhler.
4.19f	D	The bees were conditioned to the colour blue and had colour vision because they could discriminate between blue and shades of grey that would have been indistinguishable from blue if they had only monochrome vision.
4.20f	B	A bridge is used in animal training to indicate that a reward is coming. It is useful when it is not possible to give the reward immediately after the desired behaviour has been performed. It is also called an event marker.
4.1i	A	Trial-and-error learning and operant conditioning are essentially the same thing.
4.2i	B	This is a learning curve. The figure shows the decrease in the time taken to complete the maze with an increasing number of trials. It could also be constructed in terms of the change in the number of errors (wrong turns) made in successive trials.
4.3i	C	Semantic memory is concerned with an animal's world knowledge that she has accumulated throughout her life, rather than memories of specific individual events.
4.4i	D	All of these factors enhance the transfer of information to long-term memory.
4.5i	A	Negative reinforcement involves the removal of an unpleasant stimulus, e.g. removing restrictions from an animal when it complies with a command.
4.6i	D	Imprinting occurs during a sensitive period after birth or hatching.
4.7i	B	Müller was German and identified Müllerian mimicry; Bates was English and identified Batesian mimicry.
4.8i	C	This is probably an example of insight learning provided she had not previously seen another elephant perform this behaviour and then imitated it.

4.9i	B	The clicker and whistle are used as event markers (bridges) during training so they are second-order reinforcers. The primary rewards are the treat and the toy.
4.10i	B	This process of gradually changing behaviour is known as 'shaping'.
4.11i	B	This is sexual imprinting. It is important that young individuals imprint on their own species so that they select appropriate mates when they are older.
4.12i	D	This is extinction. It may be reversed by presenting the unconditioned stimulus with the conditioned stimulus again.
4.13i	A	Thorndike studied trial-and-error learning by putting a cat inside a locked box from which it could escape only if it manipulated the locking mechanism correctly.
4.14i	C	C is correct. Once the animal performs the correct behaviour the clicker is used to indicate that a reward will follow.
4.15i	D	Imprinting occurs at an early stage in an animal's life so it is phase-sensitive.
4.16i	B	This is episodic memory: memory of everyday events that occurred at specific times and places.
4.17i	A	When an animal is repeatedly exposed to a stimulus it may become sensitised to it. In addition to continuing to respond to a repeated stimulus there may be an enhancement of the response to a whole group of related stimuli. In habituation the animal learns to ignore stimuli that are irrelevant.
4.18i	D	Satiation is the temporary loss of effectiveness resulting from the repeated presentation of a reinforcer.
4.19i	D	In Müllerian mimicry similar noxious species are protected by virtue of their morphological similarities. This phenomenon was first identified in tropical butterflies.
4.20i	C	The bell began as a neutral stimulus. Once it became associated with the presence of food it became the conditioned stimulus.
4.1a	B	Ribonucleic acid is correct.
4.2a	A	This is reinforcement that is applied after a constant amount of time.
4.3a	D	Correct performance of the behaviour is rewarded only after it has been performed a specific number of times.
4.4a	B	B is correct. Note that negative reinforcement is not punishment but the removal of an unpleasant stimulus.
4.5a	B	The process of chaining breaks a task down into several smaller steps so that it can be learned in stages.
4.6a	C	These dogs were all taught to drive a car.
4.7a	C	This behaviour is known as 'isothermal tracking'. An isotherm is a line joining points of equal temperature.

4.8a	D	This was recorded in Japanese macaques.
4.9a	A	This is social facilitation.
4.10a	C	Young birds listen passively in the sensitive period after hatching and then, as juveniles, form a template and then a subsong. The crystallised final song appears in the mature adult.
4.11a	C	African elephants, dogs and African giant pouched rats have all been trained to detect the explosive TNT.
4.12a	A	The modification of the growth of this plant based on its previous experiences suggests it has a type of memory.
4.13a	D	This is taste aversion otherwise known as the Garcia effect, named after John Garcia. The distracters are fictitious in this context.
4.14a	A	This study involved breeding separate strains of 'bright' and 'dull' rats based on their performance in a maze to show that the ability to learn the maze was inherited. This was an early demonstration that behaviour has a genetic component.
4.15a	C	The chicken learned to associate the sound with the appearance of a worm more quickly when the time between the sound and the appearance of the worm was shorter (10 seconds) than when it was longer (60 seconds). However, when the sound and the appearance of the worm occurred simultaneously the chicken was slow to make the association.
4.16a	A	The food provided for correct behaviour was positive reinforcement; the water spray for incorrect behaviour was positive punishment.
4.17a	D	Social learning in fishes has been demonstrated experimentally in all four contexts.
4.18a	C	The single-celled Paramecium has been shown to learn as a result of Pavlovian conditioning.
4.19a	A	The hippocampus in the brain is important in the formation of the map-like memories needed for spatial navigation.
4.20a	B	This phenomenon is known as interference.

Chapter 5

5.1f	B	Dispersal is generally defined as the movement of animals from their place of birth.
5.2f	D	The range of a species is the geographical area over which it occurs. Within this range an individual will have a home range and inside this may have a territory, depending upon the species.
5.3f	C	Compass is correct. A theodolite is used for measuring angles; a GPS is a global positioning system that determines position from satellite information; a sextant is an instrument used to determine the angle between the horizon and the Sun or other celestial body.

5.4f	D	A homing pigeon can be moved away from where it is normally kept and it will find its way home.
5.5f	A	Common wombats produce faeces that are cubical. The distracters are other marsupials.
5.6f	B	Continental drift has resulted in Ascension Island and Brazil moving apart over geological time thereby gradually increasing the distance the turtles have had to swim.
5.7f	C	The turtles lay their eggs on Ascension Island.
5.8f	A	The fish used were coho salmon (*Oncorhynchus kisutch*).
5.9f	D	Some species within all of these taxa are able to echolocate.
5.10f	A	The territories held by members of the same species may vary in size between individuals and over time.
5.11f	A	A territory is a defended area located within a home range.
5.12f	D	A celestial compass uses the stars, the Sun or the Moon. Skylight polarisation patterns can be used to detect the position of the Sun when it is obscured by cloud.
5.13f	C	Humpback whales utilise feeding grounds in polar regions and breed in tropical waters.
5.14f	C	The purpose of territoriality is to defend resources of various kinds.
5.15f	A	Polyterritoriality is the holding of more than one territory during a single breeding season.
5.16f	B	A midden is a pile of dung.
5.17f	C	Of the species listed, only red squirrels do not perform a handstand while scent marking on vertical surfaces such as tree trunks.
5.18f	B	In seasonal monogamy individuals live separately outside the breeding season, possibly in a flock, and form pairs only during the breeding season.
5.19f	D	Artic terns have the longest annual migrations: a round trip of around 35,000 km each year between the Arctic and the Antarctic.
5.20f	A	An irruption is a sudden increase in numbers.
5.1i	B	Territoriality is favoured if the benefits of holding a territory outweigh the costs of defending it. Costs are incurred when an individual has to patrol the territory, exclude intruders and possibly engage in combat. Benefits accrue in the form of access to food, shelter, mates and breeding sites.
5.2i	C	The butterflies migrate from Canada and the United States to central Mexico, i.e. north-south not west-east.
5.3i	D	This is natal philopatry, although the term is sometimes used to describe repeatedly returning to the same place to breed even if the animal was not born there.
5.4i	B	Lemmings migrate away from centres of high density when populations increase and density rises resulting in food scarcity.

5.5i	A	Anadromous fish migrate from seawater to freshwater to spawn. Catadromous fish migrate from freshwater to seawater to spawn. Diadromous fish are capable of living in, and migrating between, fresh and salt water. Endogenous means originating from within an organism.
5.6i	C	A time-compensated compass is a compass that takes into account the changes in position of celestial bodies at different times of the day.
5.7i	B	Sonar affects cetaceans (dolphin, whales and their relatives) because they use echolocation to navigate.
5.8i	A	This is a dewlap and is used for communication. Figure 5.1 shows the dewlap fully extended.
5.9i	B	A single resource territory is one that is defended for a single purpose, e.g. food resources or mating sites.
5.10i	C	As small mammal density increases the feeding territory size of birds of prey would decrease as their food is easier to find when small mammal densities are high.
5.11i	C	This is because the arrangement of the landmarks was important rather than the landmarks themselves, so the wasp would look for the shape of a circle rather than a group of cones.
5.12i	B	A magnetic compass was first discovered in homing pigeons.
5.13i	D	The oribi possesses preorbital scent glands and uses them to mark vegetation.
5.14i	A	An animal that orientates its body into the flow of a water current is exhibiting positive rheotaxis.
5.15i	D	Polarised light vibrates in one plane and can help some species to locate the Sun when it is obscured.
5.16i	C	The Midwest flyway is fictitious.
5.17i	B	Birds migrate so that they can balance their energy intake with their energy expenditure. If they were to remain in an area as temperature fell and food became scarce they would be using more energy to keep their body temperature up at a time when less energy was available from food. It is better to migrate to a place where food is plentiful and the temperature is higher.
5.18i	A	Pharotaxis is the use of landmarks in navigation.
5.19i	C	This allows some animals to determine the position of the Sun when it is obscured.
5.20i	C	The direction of flight is controlled by seasonal hormonal changes.
5.1a	D	Birds use all of these methods.
5.2a	D	This is the 'dear enemy effect' and allows individuals to reduce the amount of time and energy they spend on the defence of their territories against neighbours they know.

5.3a	C	A zeitgeber is a cue from the environment about the passage of time. It is a German word that literally means 'time-giver'.
5.4a	A	This system is known as Argos. Argos was a city in ancient Greece. The distracters are the names of other places in Greece.
5.5a	D	Westerly breezes help warblers reach islands in the Atlantic and the Caribbean, not easterly breezes.
5.6a	C	Birds that switch from being migratory to resident are usually older individuals, not younger.
5.7a	C	The territory owner always 'wins' regardless of the length of time the territory has been held.
5.8a	D	This behaviour is called site fidelity.
5.9a	D	Species in all three taxa possess a magnetic sense.
5.10a	B	The amount of access the male has had to the female determines the amount of parental effort he will expend. The more exclusive his access the more likely he is to be the father of the chicks.
5.11a	D	There is no evidence in the graph that can be used to determine which birds won the fights.
5.12a	D	Rainfall and plant nutrition gradients are the most important factors affecting this species' migratory behaviour.
5.13a	C	Burt discussed home range in mammals.
5.14a	B	Domestic cats and cheetahs both use time-plan spacing to avoid conspecifics.
5.15a	C	After hatching these turtles use visual cues to find the sea initially followed by information about the orientation of the waves. Thereafter they use magnetic orientation and subsequently wave direction.
5.16a	B	The minimum convex polygon method encloses all of the locations and creates the smallest polygon in which none of the internal angles is greater than 180 degrees.
5.17a	D	This was written by Robin Baker who worked at the University of Manchester, England, for many years.
5.18a	C	This occurred when territories were stable in the middle of the breeding season.
5.19a	C	The brass bars were not magnetic so had no effect on the magnetic compass of the birds.
5.20a	D	All of these hormones can influence migration.

Chapter 6

| 6.1f | C | Cognitive ethology is concerned with cognitive processes as they occur in nature rather than in the laboratory. |

6.2f	B	*Clever Hans* was a horse whose owner appeared to have taught him to count. An investigation in 1907 found that he was receiving cues from his trainer.
6.3f	D	D is correct.
6.4f	C	An animal that is capable of mental state attribution has the cognitive capacity to reflect on its own mental state and the mental state of others.
6.5f	A	A lexigram is a picture or symbol that represents a word.
6.6f	A	Gordon G. Gallop Jr is an American psychologist who developed the mirror test for self-recognition. The distracters are the names of scientists who have worked on primate behaviour.
6.7f	D	Some species intentionally apply scent marks on top of the scent marks of others of their species.
6.8f	C	Stridulation is the process by which sounds are produced by rubbing together certain body parts. Insects possess stridulatory organs that they rub together. This is analogous to the dragging of a needle across a vinyl record to produce a sound.
6.9f	B	This is decoding. The distracters are other terms related to the process of coding and decoding information.
6.10f	B	In mammals sleep consists of active sleep (REM – rapid eye movement – sleep) and quiet sleep (NREM – non-rapid eye movement – sleep).
6.11f	A	Pheromones occur in a wide range of taxa including ciliates and vertebrates.
6.12f	C	Urine washing occurs in capuchin monkeys.
6.13f	D	The first pheromone studied was obtained from the Chinese silk moth.
6.14f	A	Foot-flagging is a communication method used by some anurans (frogs).
6.15f	C	The vomeronasal organ detects odours.
6.16f	B	B is correct. The distracters are the same interpretations matched with incorrect images.
6.17f	C	When vervet monkeys hear the 'snake' alarm call they stand upright and look at the ground as this is where they would expect to see a snake.
6.18f	B	The 'round dance' indicates the proximity of food; the waggle dance coveys information about the distance to food and the direction.
6.19f	D	The pygmy chimpanzee (bonobo) uses an elaborate system of hand gestures to communicate with conspecifics. The distracters are other primates.
6.20f	B	Each dolphin has its own 'signature whistle' from which it can be identified by other dolphins.

6.1i	C	Köhler conducted many laboratory studies on the cognitive abilities of chimpanzees
6.2i	A	*Kanzi* was a male bonobo (*Pan paniscus*).
6.3i	C	Dialects have been found in populations of some cetaceans species. An accent is a distinctive way of pronouncing a language, often related to a particular area; a sociolect is the dialect of a specific social class; an idiolect refers to the speech habits of a particular person.
6.4i	B	African elephants create seismic waves that travel long distances.
6.5i	A	*Washoe* was taught American sign language.
6.6i	C	Some researchers call structures such an anvils, bait used to attract prey and the twigs in the bower built by a bowerbird 'proto-tools' as they are immobile and not held or manipulated by the animal.
6.7i	D	Honest signals transmit reliable information from one individual to another.
6.8i	A	A sound spectrograph produces a visual representation of a sound.
6.9i	B	A cognitive map is a mental representation of information relating to locations.
6.10i	C	In the 'waggle dance' the bee walks a path that resembles a figure '8'.
6.11i	A	This is the theory of mind and is sometimes known as 'mind-reading'.
6.12i	D	*Washoe* was observed teaching a younger chimpanzee *Loulis* American sign language.
6.13i	B	Cetaceans communicate using click trains or codas.
6.14i	C	Old World monkeys use lip smacking for communication.
6.15i	D	All of these are examples of tool use in birds.
6.16i	D	All of these methods of communication are used by African elephants.
6.17i	A	This is the handicap principle. Costly signals must be reliable. The distracters are fictitious in this context.
6.18i	B	This is antiphonal calling whereby two individuals make calls in turn.
6.19i	A	This is a deimatic display or startle display. 'Deimatic' means 'to frighten'.
6.20i	D	Dreaming occurs during active sleep not quiet sleep. Figure 6.2 is a sleeping squirrel monkey (*Saimiri*).
6.1a	B	The red spot test (mark test) is used to test animals for self-recognition. The researcher surreptitiously places a red spot on the face of the animal and records whether or not it touches the spot more than it touches other parts of its face.
6.2a	A	Darcin is a male sex pheromone found in mice that attracts females. It was named after the character 'Darcy' in Jane Austen's book *Pride and Prejudice*.

6.3a	D	These pheromones allow males to identify previous opponents and thereby help to maintain dominance hierarchies. They are also important in mate choice and mating behaviour. Males that win fights are more attractive to females than those that do not.
6.4a	C	Vocal clans is correct. The distracters are fictitious in this context.
6.5a	B	Sifakas are lemurs from the family Indriidae.
6.6a	A	Most alarm calls are genuine and ignoring an alarm call could be fatal.
6.7a	C	Ravens (*Corvus corax*) hold stones and other objects in their beaks in a referential gesture called 'showing'.
6.8a	B	Some, but not all, chimpanzees appear to recognise themselves in a mirror using this test.
6.9a	A	Luciferin is oxidised by the enzyme luciferase to produce light. The distracters are all receptors/ pigments found in cells in the retinas of various taxa.
6.10a	D	This device is being used to investigate trial-and-error learning.
6.11a	D	All of these animals practice antiphonal calling, i.e. vocalise in turn.
6.12a	B	This is the 'audience effect' or the 'bystander effect'. The distracters are fictitious in this context.
6.13a	C	This is metacommunication: a secondary communication indicating how subsequent information is to be interpreted.
6.14a	A	This is an abdominal gland that produces chemicals that have a number of functions including as a sex attractant in Hymenoptera.
6.15a	B	This is called eavesdropping.
6.16a	D	New Caledonian crows are able to assemble compound tools. They can construct a long stick from shorter elements that can be joined together.
6.17a	B	B is incorrect. Low frequency sounds travel further than high frequency sounds.
6.18a	A	These animals were Asian elephants.
6.19a	D	All of these are features of a referential gesture.
6.20a	C	This rodent makes sounds by hitting the ceiling of its tunnel with its head.

Chapter 7

7.1f	D	All three species engage in food-begging whereby adults regurgitate food for their young.
7.2f	C	In fission-fusion societies the individuals in a group change with time.

7.3f	B	Naked mole rats are eusocial mammals. Eusocial animals live in colonies in which there is a division of labour among animals that belong to different castes which are behaviourally distinct, including non-reproductive groups. Young are reared in broods where individual adults help to rear the young of others.
7.4f	D	Vero Wynne Edwards argued that many species have evolved adaptive population-regulatory mechanisms whereby individuals communicate information about population density using epideictic displays. His ideas received a great deal of criticism from other zoologists.
7.5f	B	The behaviours described are typical submissive behaviours in mammals. Some have been ritualised from behaviours that originally evolved for a different purpose, e.g. presenting the buttocks is a behaviour associated with receptive female primates.
7.6f	A	Red deer hinds incur greater costs in suckling males than females with a consequent effect on the mother's survival.
7.7f	C	Kin selection considers the role of altruistic behaviour in increasing the number of genes an individual leaves to the next generation by assisting relatives to rear their young.
7.8f	D	Giraffes live in fission-fusion social groups where associations are stronger if young are present and these young are reared in nursery groups.
7.9f	B	These animals 'rut' at certain times of the year.
7.10f	D	This behaviour is mate guarding and has been recorded in a number of species including African elephants.
7.11f	C	Parental investment is the cost to parents of rearing their young.
7.12f	B	Polygyny refers to the many (poly) females (gyny) mated by each male.
7.13f	A	Gemsbok with symmetrical horns have greater genetic fitness than those with asymmetrical horns.
7.14f	C	These groups are known as castes. The distracters are fictitious in this context.
7.15f	D	These areas are called 'leks'. The distracters are fictitious in this context.
7.16f	B	'Allomother' means 'other' or 'different' mother.
7.17f	B	A brood parasite, such as a cuckoo, hatches in the nest of a different species and is fed and cared for by individuals of this species.
7.18f	A	Lyrebird birds court females on circular mounds of bare soil on the floor of the forest.
7.19f	B	A male dunnock pecks at the female's cloaca before mating with her to induce her to eject the sperm of any male with whom she has previously mated.

7.20f	A	In a linear hierarchy each individual has its place in a linear sequence so in this example individual G is dominant to B, F and D, Individual B is dominant to F and D, and F is dominant to D.
7.1i	C	Selfish herd theory is correct. The distracters are fictitious in this context.
7.2i	A	This is the mother hypothesis.
7.3i	D	In this species there is a hierarchy of males and a separate hierarchy of females.
7.4i	A	Humpback whales produce bubble nets in which they catch food organisms such as small fishes or krill. The 'net' is created by exhaling through the blowhole. Bubbles temporarily trap the fish while the whales move in to feed.
7.5i	B	Appeasement behaviour serves to prevent attack and often involves lowering the head and body.
7.6i	D	All three scenarios would make the ability to change sex advantageous.
7.7i	D	Evolutionarily stable strategy is correct. The distracters are fictitious in this context.
7.8i	A	In olive baboon societies a young female 'inherits' her rank from her mother.
7.9i	A	A 'faeder' is a male ruff whose plumage makes it look like a female. This makes it easier to mate with a female because it does not attract the attention of other males.
7.10i	B	This behaviour allows males to assess each other's strength. Roaring is energetically expensive. Stronger males are able to roar for longer than weaker individuals.
7.11i	D	All of these animals have 'helpers at the nest'.
7.12i	C	This is 0.25 (uncle to niece) x 0.50 (mother to son) = 0.125.
7.13i	A	If a male is to invest time in rearing young it is important that he has a high degree of confidence that he is their father. This is known as parental certainty.
7.14i	C	This gift is given to the female so that the male can mate with her while she is eating. In the absence of the food she might eat him instead.
7.15i	D	Individuals in these taxa live in societies where the knowledge held by older individuals is an important resource for others.
7.16i	C	In resource defence polygyny a male has more than one mate and monopolises the resources the females need for breeding success.
7.17i	C	Inclusive fitness consists of the fitness of an individual that is derived from direct reproduction (direct fitness) taken together with the fitness derived from helping close relatives to survive and reproduce (indirect fitness).

7.18i	B	This is an asymmetric contest. The distracters are fictitious in this context.
7.19i	D	All of these taxa contain species that engage in brood parasitism.
7.20i	C	The Fraser Darling effect is the synchronisation of breeding that is compressed into a short breeding season found in some social animals to increase the chance of survival of individual offspring.
7.1a	C	This is a definition of evolutionary game theory. The distracters are fictitious in this context.
7.2a	B	This is scramble competition polygyny. Polygyny refers to 'many females'.
7.3a	D	This is the good sperm hypothesis. The distracters are fictitious in this context.
7.4a	D	This is a social behaviour that occurs after mating in which the pair and others of the social group gather together briefly, vocalising loudly.
7.5a	A	This is reciprocal altruism as both animals perform altruistic acts and both benefit from the relationship.
7.6a	A	A game matrix would be a 2x2 matrix showing the costs and benefits of chasing large and small prey at fast and slow speeds and would allow a determination of the best feeding strategy. Catching a large prey provides more energy than catching a small prey. Running fast uses more energy than running slowly.
7.7a	C	Hyenas are promiscuous and form no long-term pair bonds.
7.8a	B	The prisoners' dilemma is a game used in game theory to analyse when it is in the interests of two individuals to cooperate and when it is not.
7.9a	B	This is Hamilton's rule, named after the geneticist W. D. Hamilton. The distracters are fictitious in this context.
7.10a	C	C provides the best protection from predators as it is completely surrounded by other individuals making it less likely to be attacked.
7.11a	D	The concept of a pecking order was derived from studies of domestic chickens.
7.12a	B	Mafia hypothesis is correct. The distracters are fictitious in this context.
7.13a	C	This is a sneaky mating. The distracters are fictitious in this context.
7.14a	C	The presence of broken shells would indicate the presence of chicks (food) and make the offspring of ground-nesting birds very susceptible to predation. Cliff-nesting birds are less susceptible to predation because of the relative inaccessibility of their nests.
7.15a	C	C is correct. Most mobbers (77%) are breeding adults with young to defend so it appears that the behaviour is a form of parental care.

7.16a	B	Some birds produce two chicks, the older of which kills the younger in a behaviour known as obligate siblicide. The insurance egg hypothesis explains the production of the second egg as insurance against the failure of the first egg.
7.17a	C	If the rate at which costs are accrued during the contest are identical for A and B, A will give up if the value it places on the resource is lower than that placed on the resource by B. If the value of the resource is the same for both A and B, A will give up if the costs it incurs in fighting are higher than those incurred by B.
7.18a	B	The value of resources to a contestant is best described as the increase in Darwinian fitness that they confer.
7.19a	A	The animals must have a long lifespan to allow time for reciprocation. If an animal has a short lifespan it could act altruistically towards another individual and then die before this could be repaid.
7.20a	C	This is optimality theory. Natural selection favours the variant of a behaviour which provides the greatest net benefit.

Chapter 8

8.1f	D	A condition score is designed to assess and monitor body condition from an animal's physical appearance, e.g. whether or not ribs are visible, the amount of fat present, etc. Body condition scores are widely used to assess the welfare of farm animals but have also been used to study free-ranging animals and those kept as companion animals and in zoos.
8.2f	C	This is a longitudinal study: the same subjects are studied over a long period of time.
8.3f	B	This is focal sampling. The animal being followed is the focal animal.
8.4f	B	High-speed photography is correct. Consecutive images are recorded at a high frame rate and played back at a much lower frame rate thereby producing a slow motion sequence. If an event lasting 10 seconds is recorded at 100 frames per second and played back at 25 frames per second the event will be slowed down and appear to last 40 seconds.
8.5f	C	This question is part of a self-administered questionnaire – because the respondent must fill in a form – and is closed because the respondent must chose between one of four responses and cannot respond in any other way, e.g. by creating a fifth option or by adding a written response.
8.6f	A	*Ad libitum* sampling allows the recording of every instance of a behaviour that is relatively rare, for example instances of mating. The term literally means 'as often as desired'.

8.7f	D	All of these methods are acceptable but some scientists believe it is inappropriate to use names normally associated with humans. The name is merely a label and it should not really matter whether we call an individual '23', 'K', 'red' or 'Jack'.
8.8f	B	This is a preference test. The distracters are fictitious in this context.
8.9f	A	The Open Field Maze was designed to measure emotionality in rodents. It consists of an area enclosed by a wall that is sufficiently high to prevent escape and large enough for the subject to experience the feeling of openness when in the centre of the maze. Measurements are made of total distance covered by the subject (ambulatory distance), tendency to remain near walls, and the number of faecal boli deposited.
8.10f	A	The 'five domains' model of animal welfare was proposed by D. J. Mellor and C. S. W. Reid in 1994. Note that this is not a reference to the 'five *freedoms*' model. The distracters are fictitious but contain the names of well-known authors in the field of animal welfare.
8.11f	D	'Pithing' refers to killing an animal by severing its spinal cord. It may be used to kill or immobilise an animal.
8.12f	A	Biting and copulation are events as they are of relatively short duration. Feeding, sleeping and suckling are ongoing states.
8.13f	C	The weights were replicates as they related to different individuals weighed at the same age.
8.14f	A	A Skinner box is named after B. F. Skinner and is used to study operant conditioning, a type of learning.
8.15f	B	The term 'actograph' may refer to a device that monitors and records a number of types of animal movement. It may also refer to an obsolete device for recording muscle movements.
8.16f	D	A parabolic reflector is a dish used to 'collect' sound waves and focus them towards a microphone in a manner similar to that in which a satellite dish receives signals from a satellite. A parabolic reflector allows sound to be recorded that is coming from a specific direction.
8.17f	C	This is a choice chamber. It looks similar to a Petri dish used in microbiology but is larger.
8.18f	C	Spatial proximity loggers are electronic devices that can be worn by individual animals and will record when they have been near to others wearing a similar device.
8.19f	B	A sociogram can be used to show the social relationships within a group of animals of any size.
8.20f	B	A cathode ray oscilloscope will produce a visual representation of the magnitude and speed of transmission of a nerve impulse.

8.1i	D	There are 3x12=36 recordings possible if they are made every 5 minutes for 3 hours. The animal was out of sight for 8 recordings so 28 behaviour recordings were made. The rhinoceros was feeding on 23/28 recordings = 82.1%.
8.2i	C	This is instantaneous scan sampling. The group was scanned every 5 minutes and the behaviour of each at that instant was recorded.
8.3i	A	This is a Monte Carlo simulation. The outcome depends on chance (like gambling in a casino in Monte Carlo) and is different each time the simulation is run.
8.4i	B	This apparatus was used to study imprinting in ducklings. The object 'X' was an adult duck decoy.
8.5i	C	Latency is the time between a stimulus and a response.
8.6i	D	This is a mathematical model.
8.7i	B	The figure shows models made to represent mothers to infant monkeys. The young monkeys were observed clinging to the cloth 'mother' while reaching across to the wire 'mother' to feed from the feeding bottle.
8.8i	C	DL_{50} refers to the lethal dose required to kill 50% of the test animals. It has been widely used to test the toxicity of various chemicals and criticised because it results in the deaths of a large number of animals.
8.9i	D	Eye temperature can be used to assess animal welfare. For example, there is a positive correlation between eye temperature and heart rate variability (indicating stress) in horses.
8.10i	B	Animal welfare cannot be directly measured.
8.11i	B	These are 'honest' signals because they accurately reflect the suffering of the animals, and the signals are 'inter-species' because they are transmitted by piglets and received by humans.
8.12i	A	These are consumer demand tests and objectively measure an animal's motivation to obtain resources.
8.13i	B	A sociability chamber has three chambers arranged in sequence. It is used to compare the test animal's response to an animal it has encountered previously and an 'intruder'.
8.14i	D	A Barnes maze is used to study spatial learning and memory in the laboratory. See Fig. 8.7.
8.15i	B	The list in B is correct. The 'five domains' model has a strong focus on mental wellbeing and positive experiences whereas the 'five freedoms' model focusses on the absence of negative experiences.

8.16i	A	This is a reference to the horse called *Clever Hans* who appeared to be able to perform a number of cognitive tasks but was in fact responding to cues from his owner. The presence of unintentional cues may affect the outcome of poorly designed cognition experiments where the subject can see the researcher.
8.17i	B	Measuring these characteristics allows us to 'quantify' the personality of an animal.
8.18i	B	This is one-zero sampling. During the sampling period a particular behaviour either occurs or it does not occur. The number of occurrences in each sampling period is not recorded.
8.19i	D	The variable on the *y*-axis is dependent (because its value depends upon the time of day) and continuous (because it can take any value between certain limits).
8.20i	D	The purpose was to ensure that each observer was recording the various behaviours in the ethogram in the same way so that data from all four observers could be pooled.
8.1a	C	This phenomenon is called 'observer drift'. For example, if all types of stereotypic behaviour are pooled in a single category of behaviour this category may be unconsciously expanded with time as rare stereotypies are observed late in the study that might have been overlooked earlier.
8.2a	A	It is perfectly legitimate to include a behaviour in an ethogram if its purpose is unknown. By recording the occurrence of an unusual behaviour it may later be possible to correlate this with an environmental variable that helps to elucidate its purpose.
8.3a	A	The number of different dyads that could occur within a group is calculated as (Number of individuals x (Number of individuals − 1))/2. In this case (4x3)/2 = 6.
8.4a	C	If the index of association is 1.0 the two animals are always seen together; if it is 0 they are never seen together; if it is 0.5 they are seen together 50% of the time. If the index is 0.23 they are seen together 23% of the time, i.e. fewer times than they have been seen apart.
8.5a	D	The highest value the index can have is 1 not 100. An index of 1 indicates that the two animals are always seen together.
8.6a	D	A payoff matrix is used in game theory. It shows the various possible strategies and outcomes, and allows us to predict a strategy.
8.7a	A	J. Maynard Smith was a theoretical ecologist. He defined a 'strategy' as a behavioural phenotype.
8.8a	B	This is a Barnes maze. A mouse is placed in the centre of the circular enclosure and can only escape through one of a series of holes around the periphery. The mouse learns the position of the escape hole with reference to the spatial cues at the edge of the maze. The maze was first developed by Dr Carol Barnes.

8.9a	C	Bateson's cube is named after the zoologist Patrick Bateson. Its three dimensions are (1) the quality of research, (2) the certainty of medical benefit, (3) the amount of animal suffering.
8.10a	B	This is a confounding variable. For example, the age of an animal may affect the result of an experiment and may not have been taken into account in the experimental design.
8.11a	D	All three features listed are important in the design of an Open Field Maze.
8.12a	A	The Dawkins Organ was an electronic device invented by the zoologist Richard Dawkins for recording behaviour events on a timescale so that they could be analysed with a computer.
8.13a	D	The 'five domains model' does all of these things. The 'five freedoms model' focussed on the removal of harmful experiences.
8.14a	C	This paper is historically important because it drew together the various sampling methods being used to study social behaviours at that time.
8.15a	C	Clock-shifting is the phase shifting of an animal's endogenous clock. If an animal has inaccurate information about the time it cannot correctly compensate for the changing position of the Sun at different times of the day during navigation.
8.16a	B	This is a shuttle box and it is used in the study of avoidance learning.
8.17a	D	'Protected animals' under this legislation are defined as living vertebrates, other than man, and living cephalopods (e.g. octopuses and squids).
8.18a	B	A 'regulated procedure' under this legislation is essentially anything that causes more pain or distress than a needle inserted by a veterinary surgeon.
8.19a	B	The graph indicates that the pigs were prepared to work harder for access to sand than for access to conspecifics (social contact).
8.20a	A	This is a negative correlation because as the work required to gain access to other pigs increased the number of rewards decreased.

Chapter 9

9.1f	A	The term 'welfare' can only relate to a single individual. Populations and species are made of individuals.
9.2f	B	B is correct. Note, this question refers to the 'five freedoms' and not the 'five domains' of animal welfare.
9.3f	C	The nickname 'Humanity Dick' refers to Richard Martin MP who was a founding member of the Royal Society for the Prevention of Cruelty to Animals (RSPCA) and responsible for passing the first law in the world to protect animals from cruelty in 1822 (Martin's Act).

9.4f	A	Descartes denied that animals could feel pain and claimed that they did not possess a mind.
9.5f	C	Primatt was an English clergyman. This book is a very early work on animal cruelty.
9.6f	B	This is tail docking. The practice is banned by law in many countries if done for cosmetic purposes.
9.7f	B	It is not necessary for animals to be assigned moral rights in order to achieve good animal welfare.
9.8f	A	This is environmental enrichment that requires the sea lion to work for its food.
9.9f	C	There is evidence that stereotypic behaviour may be associated with physical changes to the brain.
9.10f	D	The five freedoms were established by the *Report of the Technical Committee to Enquire into the Welfare of Animals kept under Intensive Livestock Husbandry Conditions* published in 1965.
9.11f	D	Several species of bears are kept on 'bile farms' in a number of Asian countries, for example, China and Vietnam.
9.12f	C	Hunger is a normal experience and motivates animals to search for and acquire food.
9.13f	B	Hip dysplasia – dislocation of the hip joint – is common in the German shepherd.
9.14f	D	Mastitis is inflammation of the breast tissue.
9.15f	B	Polled cattle have had their horns removed. Horns may be removed for safety reasons.
9.16f	A	Distress is a state resulting from an animal's inability to cope with stress.
9.17f	A	A leghold trap catches an animal (usually a large mammal) using metal jaws which close around its leg when sprung. The trap is usually tethered to a tree or other immovable structure so that the animal cannot escape.
9.18f	D	All of these animals have been hunted for sport in the United Kingdom using dogs. This practice was banned in England and Wales by the Hunting Act 2004 and in Scotland by the Protection of Wild Mammals (Scotland) Act 2002. Wild boar are no longer part of the native fauna but some farm escapes occur at various locations.
9.19f	C	The five freedoms were established in 1965 as a result of the *Report of the Technical Committee to Enquire into the Welfare of Animals kept under Intensive Livestock Husbandry Conditions*.
9.20f	C	This reported linked violence towards children with violence towards animals.
9.1i	B	'Canned hunting' is particularly associated with the shooting of lions held within fenced enclosures.

9.2i	D	This condition is associated with pugs and causes breathing difficulties. This breed has a distinctive short-muzzled face.
9.3i	A	The wool attached to this skin in sheep can become covered in faeces. This attracts flies that lay their eggs under the skin. The condition can be fatal.
9.4i	A	In overcrowded indoor conditions chickens may aggregate in large densities leading to smothering of some individuals.
9.5i	D	Pinioning is a method of flight restraint achieved by the surgical removal of part of the wing.
9.6i	B	Aphagia is the inability to swallow.
9.7i	C	These animals cannot raise themselves from the ground or walk. This may be due to broken legs, a fractured vertebral column, severed tendons or ligaments, paralysis or some other reason.
9.8i	B	These are battery cages.
9.9i	B	Farrowing crates restrict the movement of the sow so that she does not crush her piglets.
9.10i	C	SPIDER stands for Setting goals, Planning, Implementing, Documenting, Evaluating and Readjusting.
9.11i	D	Increased lamb survival has been achieved by controlling disease using vaccines and anthelmintics (for parasites).
9.12i	D	Red foxes were deliberately introduced into Australia specifically so they could be hunted.
9.13i	A	A captive bolt is used to stun livestock to induce unconsciousness prior to slaughter. Modern devices resemble a gun. The bolt is a metal rod that is propelled forward by a spring or compressed air. It may or may not destroy part of the brain, depending upon the design.
9.14i	C	This process is called counterconditioning.
9.15i	B	FCM stands for faecal cortisol metabolites. These substances are indicators of stress.
9.16i	B	Many farmed sheep are kept indoors on floors made of wooden slats.
9.17i	A	A gavage tube is a flexible tube that is pushed into the stomach of an animal so that it may be force fed or to administer drugs.
9.18i	D	The farmers train pig-tailed macaques to collect coconuts. When they are not working they are usually kept chained.
9.19i	B	Poor stockmanship, for example moving animals too quickly, directly affects the frequency of lameness in dairy cattle.
9.20i	A	This increase is most likely to have been the result of damage to the housing stock in Liverpool and the consequent social deprivation. In WWII Liverpool was the most bombed area of the United Kingdom after London.

9.1a	B	In a free-flow robotic milking system the cows voluntarily present themselves for milking. This eliminates the stress experienced when animals are moved by humans. In addition, robotic systems use electronic tags to identify each animal so that the frequency with which she visits the milking machine and the quantity of milk delivered are measured. This data supplements other health and welfare data collected by the farmer.
9.2a	C	This is a housing system used for pigs.
9.3a	D	A 'priest' is a small mallet used to kill fish and other small game animals.
9.4a	D	Over 1000 inherited disorders are known in domestic dogs.
9.5a	C	Feline herpes virus 1 and feline calicivirus cause cat flu.
9.6a	A	This practice is known a 'soring'. It is still used in parts of the United States despite having been banned under the Horse Protection Act of 1970.
9.7a	B	Frances Cobb was an Irish pioneer of animal rights who founded this society.
9.8a	D	This system was introduced to improve conditions for pigs.
9.9a	A	The morbidity rate measures the rate of incidence of disease and should not be confused with the mortality rate which is the death rate.
9.10a	D	Padded, offset and laminated are all types of leghold trap.
9.11a	D	All of these things are characteristics of mice kept in enriched conditions.
9.12a	A	In 2017 Ringling Bros. and Barnum and Bailey announced it was closing. The organisation was famous for it circuses and its performing animals, including a large number of elephants.
9.13a	B	These animals were all sent into space. *Kaika* was a dog; *Ham* was a chimpanzee; *Félicette* was a cat; and *Albert II* was a rhesus monkey.
9.14a	C	This test was designed to measure the efficacy of drugs designed to treat depression. The more efficacious the drug the longer the animal would continue its attempts to escape.
9.15a	A	Serotonin affects mood and sense of wellbeing. It also has effects on cognition, learning and memory.
9.16a	D	The Directive requires all of the things listed.
9.17a	B	Unwanted, lost and feral cats are the domestic animals most likely to be taken to RSPCA centres.
9.18a	D	The authors found all of these statements to be true.
9.19a	B	One quarter of the offspring of heterozygous parents will be homozygous for the condition and inherit the disease.

9.20a	A	American Humane (formerly the American Humane Society) operates this certification scheme.

Chapter 10

10.1f	B	The study of animal rights is concerned with the ethical treatment of animals so it is a branch of philosophy. This should not be confused with animal welfare science.
10.2f	C	The Greek philosopher Aristotle believed that the superiority of humans over animals was founded on the former's ability to reason.
10.3f	A	This is an aversive stimulus.
10.4f	B	Vegan is correct. A vegetarian eats only plants and may also abstain from the use of animal by-products (e.g. leather); a lacto-vegetarian excludes meat, fish, poultry and eggs from the diet; a pescatarian eats a vegetarian diet but also eats fish and other seafood.
10.5f	C	This is one way of defining sentience.
10.6f	C	Singer wrote *Animal Liberation*. This book was important in establishing the animal liberation movement. It focussed primarily on the use of animals in science and agriculture.
10.7f	B	Anthrozoology is the science that studies the relationship between humans and other animals, especially companion animals and those living on farms and in zoos.
10.8f	D	Camel wrestling is particularly popular in Turkey.
10.9f	D	The IUCN is the International Union for the Conservation of Nature and is not specifically concerned with animal suffering in captivity. CAPS is the Captive Animals' Protection Society; ADI is Animal Defenders International; HSI is Humane Society International.
10.10f	A	A is correct. The distracters are fictitious.
10.11f	B	The English philosopher Jeremy Bentham said this.
10.12f	C	PETA stands for People for the Ethical Treatment of Animals.
10.13f	C	Vernon Kisling is an expert on zoo and aquarium history.
10.14f	A	Vivisection is the practice of performing surgical procedures on live animals for the purpose of experimentation.
10.15f	A	Anthropomorphism is the tendency to attribute human characteristics to animals. This can cause difficulties in interpreting behaviour because, for example, the meaning of non-human primate facial expressions cannot be understood by comparing them with human facial expressions.
10.16f	B	Although bipedalism (the ability to walk upright on two legs) is considered a human characteristic it is irrelevant to the issue of human and animal rights.

10.17f	D	The use of the precautionary principle is widely advised by scientists with respect to the danger of exposure to certain chemicals, climate change and other environmental issues.
10.18f	C	As a general legal principle non-human primates (and other non-human animals) are not considered to be legal persons by the courts. Corporations and other organisations are considered legal persons and, in recent years, part of the Amazon rainforest in Colombia and a river in New Zealand have been granted legal personhood.
10.19f	A	The term 'person' appears in many legal documents but it is not synonymous with the term 'human'. As things that are not human can be legal persons it has been argued that this precedent should allow for the concept to be extended to non-human animals.
10.20f	B	Cockfighting and dog fighting were banned in 1835 by the Cruelty to Animals Act.
10.1i	A	This was argued by St Thomas Aquinas. The distracters are other Italian saints.
10.2i	C	PETA was founded in 1980 in the United States.
10.3i	D	This centre was founded by the theologian Revd Prof. Andrew Linzey and works in the Faculty of Theology at the University of Oxford. The distracters are other individuals interested in animal rights.
10.4i	B	This term was first coined by Richard Ryder. The distracters have all written about animal rights.
10.5i	A	Animals generally have no standing in court because they do not possess a legal personality.
10.6i	C	Ahimsa is correct. The distracters are other Sanskrit terms.
10.7i	A	An Act to Prevent the Cruel and Improper Treatment of Cattle 1822 (Martin's Act).
10.8i	B	This is the Royal Society for the Prevention of Cruelty to Animals (RSPCA) which was founded in 1824 as the Society for the Prevention of Cruelty to Animals.
10.9i	D	Hinduism treats all of these animals as scared.
10.10i	B	This was utilitarianism, a concept championed by Jeremy Bentham.
10.11i	C	To have intrinsic value an animal merely needs to exist. Intrinsic value comes from within and is an inherent part of something.
10.12i	B	The treatment of animals as tradable resources is called commodification.
10.13i	A	Animals cannot form the type of social contract created by humans because they cannot consent to surrendering some of their freedoms.
10.14i	C	C is correct.

10.15i	A	The term 'autonoetic' refers to the capacity to be aware of one's existence in time.
10.16i	C	Draize test is correct. The distracters are fictitious in this context.
10.17i	D	The 'Three Rs' refer to the replacement of animals in testing, the reduction in the number of animals used per experiment, and the refinement of testing methodologies to improve animal welfare.
10.18i	D	Kant was a German philosopher who believed that a being deserves moral consideration because it possesses the quality of personhood. Things are not persons so do not deserve moral consideration.
10.19i	C	C is correct. The distracters are fictitious in this context.
10.20i	B	The Brown Dog Affair concerned a controversy over whether or not a dog used in a vivisection was adequately anaesthetised and the legality of the experiment. The distracters are fictitious in this context.
10.1a	D	Mary Midgley was an English philosopher with an interest in animal rights.
10.2a	D	Fishes, crustaceans and cephalopods (octopuses and their relatives) have been scientifically shown to be sentient, i.e. not insentient.
10.3a	A	Cognitive dissonance is a mental conflict that occurs when a person's beliefs and behaviours are out of alignment, for example, claiming to be vegan but using a briefcase made of leather.
10.4a	C	*Habeus corpus* is a legal concept whereby a writ requires a detained person to be brought to a court to secure his or her release unless it can be shown that the detention is lawful.
10.5a	B	This book was written by the philosopher Tom Regan. The distracters are others interested in animal rights.
10.6a	D	Tommy was a chimpanzee who was held in a cage behind a trailer sales park in New York.
10.7a	A	The subjects of this case were five wild-caught orcas (killer whales).
10.8a	A	The 13th Amendment to the US Constitution does not mention the word 'person'. Its purpose was to abolish slavery and involuntary servitude.
10.9a	B	Painism claims that the right moral action should result in reducing the pain of the individuals that suffer the most.
10.10a	C	The effective altruist movement advocates the use of evidence and reasoning to identify the ways that the lives of others can be improved.
10.11a	C	Switzerland protected the dignity of animals in its constitution in 2008.
10.12a	D	New Zealand banned the use of non-human hominids in research, testing and teaching – except where the activity is in the best interests of the animals – in its Animal Welfare Act in 1999.

10.13a	A	Human exceptionalism treats humans as exceptional because of their distinctive capabilities.
10.14a	B	These are marginal cases. They are marginal because, although these individuals are human, they lack some of the capacities that humans normally possess.
10.15a	A	*Sandra* was living in the Buenos Aires Zoo in Argentina.
10.16a	B	This case was brought on behalf of seals living in the North Sea. The distracters are fictitious.
10.17a	B	This case was an attempt to prevent the importation into the United States of African elephants from Swaziland. The case was complicated by the fact that if the animals had not been exported from Swaziland they were due to be culled. The court determined that life in US zoos was preferable to culling. Swaziland has been known as the Kingdom of Eswatini since 2018.
10.18a	C	This was the Welsh Parliament (Senedd Cymru) in the United Kingdom.
10.19a	D	This law used a wide definition of 'cattle' to include most of the livestock kept on the farms of the day.
10.20a	C	This case was brought on behalf of the bean goose (*Anser fabalis*).

References

Altmann, J. (1974) Observational study of behaviour: sampling methods. *Behaviour* 49, 227–266.

Beck, B.B. (1980) *Animal Tool Behavior. The Use and Manufacture of Tools by Animals*. Garland STMP Press, New York.

Bédécarrats, A., Chen, S., Pearce, K., Cai, D. and Glanzman, D.L. (2018) RNA from trained *Aplysia* can induce an epigenetic engram for long-term sensitization in untrained *Apylsia*. *eNeuro* 14 May 2018, 5(3) doi:10.1523/ENEURO.0038-18.2018.

Boat, B.W. (1995) The relationship between violence to children and violence to animals: An ignored link? *Journal of Interpersonal Violence* 10, 229–235.

Briefer, E., Rybak, F. and Aubin, T. (2008) When to be a dear enemy: flexible acoustic relationships of neighbouring skylarks, *Alauda arvensis*. *Animal Behaviour* 76, 1319–1325.

Burt, W.H. (1943) Territoriality and home range concepts as applied to mammals. *Journal of Mammalogy* 24, 346–352.

Croy, M.I. and Hughes R.N. (1991) The role of learning and memory in the feeding behaviour of the fifteen-spined stickleback *Spinachia spinachia* L. *Animal Behaviour* 41, 149–159.

Cullen, E. (1957) Adaptation of the kittiwake to cliff-nesting. *Ibis* 99, 275–302.

Davies, N.B. (1978) Territorial defence in the speckled wood butterfly (*Pararge aegeria*): the resident always wins. *Animal Behaviour* 26, 138–147.

Froy, H., Walling, C.A., Pemberton, J.M., Clutton-Brock, T.H. and Kruuk, L.E.B. (2016) Relative costs of offspring sex and offspring survival in a polygynous mammal. *Biology Letters* 12(9) doi:10.1098/rsbl.2016.0417.

Hensley, C. and Tallichet, S.E. (2005) Learning to be cruel?: Exploring the onset and frequency of animal cruelty. *International Journal of Offender Therapy and Comparative Criminology* 49, 37–47.

Krebs, J.R. (1982) Territorial defence in the great tits (*Parus major*): do residents always win? *Behavioral Ecology and Sociobiology* 11, 185–194.

Leslie, B.E., Meek, A.H., Kawash, G.F. and McKeown, D.B. (1994) An epidemiological investigation of pet ownership in Ontario. *The Canadian Veterinary Journal* 35, 218–222.

Mellor, D.J. and Reid, C.S.W. (1994) Concepts of animal well-being and predicting the impact of procedures on experimental animals. In: Baker, R.M., Jenkin. G. and Mellor, D.J. (Eds) *Improving the Well-being of Animals in the Research*

Environment. Australian and New Zealand Council for the Care of Animals in Research and Teaching, Glen Osmond, Australia. pp. 3–18.

Miller, C.T., Beck, K., Meade, B. and Wang, X. (2009) Antiphonal call timing in marmosets is behaviourally significant: interactive playback experiments. *Journal of Comparative Physiology A* 195, 783–789.

Narayan, E.J., Webster, K., Nicolson, V., Mucci, A. and Hero, J-M. (2013) Non-invasive evaluation of physiological stress in an iconic Australian marsupial: the Koala (*Phascolarctos cinereus*). *General and Comparative Endocrinology* 187, 39–47.

Plotnik, J.M., de Waal, F.B.M. and Reiss, D. (2006) Self-recognition in an Asian elephant. *Proceedings of the National Academy of Sciences* 103, 17053–17057.

Primatt, H. (1776) *A Dissertation on the Duty of Mercy and Sin of Cruelty to Brute Animals.* T. Cadell, London.

Rees, P.A. (1982) *The Ecology and Management of Feral Cat Colonies.* Unpublished PhD thesis, University of Bradford.

RSPCA (2019) *Trustees' Report and Accounts 2019.* Royal Society for Prevention of Cruelty to Animals. Available at https://www. rspca.org.uk (accessed 24 August 2021).

Scholz, A.T., Horrall, R.M., Cooper, J.C. and Hasler, A.D. (1976) Imprinting to chemical cues: the basis for home stream selection in salmon. *Science* 192, 1247–1249.

Shields, W.M. (1984) Barn swallow mobbing: self-defence, collateral kin defence, group defence, or parental care? *Animal Behaviour* 32, 132–148.

Smith, J.M. (1984) Game theory and the evolution of behaviour. *Behavioral and Brain Sciences* 7, 95–101.

Swartz, K.B. and Evans, S. (1991) Not all chimpanzees (*Pan troglodytes*) show self-recognition. *Primates* 32, 483–496.

Taylor, A.A. and Weary, D.M. (2000) Vocal responses of piglets to castration: identifying procedural sources of pain. *Applied Animal Behaviour Science* 70, 17–26.

Tinbergen, N. (1963) The shell menace. *Natural History* 72, 28–35.

Tryon, R.C. (1940) Genetic differences in maze-learning ability in rats. *Yearbook of the National Society for Studies in Education* 39, 111–119.

Turkington, R., Hamilton, R.S. and Gliddon, C. (1991) Within-population variation in localized and integrated responses of *Trifolium repens* to biotically patchy environments. *Oecologia* 86, 182–192.

Printed and bound by CPI Group (UK) Ltd, Croydon, CR0 4YY

29/10/2024

14582565-0002